THE BBKA GUIDE TO
BEEKEEPING

Second edition

THE BBKA GUIDE TO
BEEKEEPING

Second edition
Ivor Davis and Roger Cullum-Kenyon

B L O O M S B U R Y
LONDON · NEW DELHI · NEW YORK · SYDNEY

The British Beekeepers Association is a world leading charity supporting honey bees and promoting excellence in beekeeping.

The Association provides an extensive programme of education, resources and advice for beekeepers, together with scientific research into the honey bee to help further our understanding of this insect and the problems now facing pollinators.

For further information about the BBKA's educational, research and other initiatives please visit – www.bbka.org.uk

Bloomsbury Natural History

An imprint of Bloomsbury Publishing Plc

50 Bedford Square
London
WC1B 3DP
UK

1385 Broadway
New York
NY 10018
USA

www.bloomsbury.com

First published 2015
© Ivor Davis and Roger Cullum Kenyon, 2015
© Photos: Ivor Davis and Roger Cullum Kenyon 2015
and others as acknowledged on page ii.

British Library Cataloguing-in-Publication Data

A catalogue record for this book is available from the British Library.

ISBN: PB: 978-1-4729-2089-8
ePDF: 978-1-4729-2091-1
ePub: 978-1-4729-2090-4

10 9 8 7 6 5 4

Printed in and bound in China by RR Donnelly Printing Solutions Ltd.

To find out more about our authors and books visit www.bloomsbury.com. Here you will find extracts, author interviews, details of forthcoming events and the option to sign up for our newsletters.

Preface

I became involved in beekeeping about 25 years ago. My wife and I had always been interested in wildlife and especially in butterflies and wild orchids. But initially it was our son who heard about beekeeping and wanted to 'have a go'. This was the start of a rapid learning experience to ensure we cared for the insects we had acquired and prevented them from causing a nuisance to our neighbours.

Eventually I was elected as a trustee of the British Beekeepers' Association (BBKA). As a trustee I recognised that the only way to really understand bees and beekeeping was to spend time learning about them both from reading books and attending study groups conducted by true masters of the craft. This was matched with spending more time working with my own bees. Our number of colonies rapidly grew from just four to over 25 and I soon started to take examinations and assessments in beekeeping. I became a Master Beekeeper and then went on to attain the National Diploma in Beekeeping qualification, the highest qualification available to amateur beekeepers in the UK.

Dr Ivor Davis NDB

Following the publication of the UK Government's *'Honey bee health strategy'*, in 2008, Roger Cullum-Kenyon and I started to work together on a new project with the BBKA which had decided to develop in a comprehensive training package to support beekeeping tutors.

This guide can be used as a companion manual to Beginners' or improvers' courses. However, even if you do not wish to go on a course the book describes some of the fascination of bees and beekeeping and it is valuable reference source for many aspects of beekeeping.

Many books are available covering all aspects of bees and beekeeping and, as your interest in beekeeping develops you will undoubtedly be keen to collect a whole library on the subject. This book is supported and endorsed by the BBKA because a useful stand-alone resource for anyone wishing to understand beekeeping.

Beekeeping is a truly absorbing subject which we hope will give you as much pleasure as it has given me.

Acknowledgements

The authors would like to thank the following:

Ivor's brother Paul who read and corrected the original manuscript and the many friends and colleagues who have been generous with their photographic and illustrative skills. Their contributions to this book are greatly appreciated.

Gerry Collins for his gift of bee related images, Roger who gave up fly-fishing to photograph bees and flowers and design this book. Robin Dartington for details of his hive, Ann Holderness for her queen mating image, Julian Routh and Jo Widdicombe for images of bees and equipment. Defra for permission to use images of pests and diseases and Omlet for the Beehaus image.

We also wish to express our thanks to Lisa Thomas our tireless editor at Bloomsbury Natural History and Bill Stevens of National Bee Supplies, for lending us hives and many other bee-keeping items to photograph as well as Martyn Norsworthy for the studio images.

I am thankful for all the help and encouragement given to me by so many people who are the cream of beekeeping in the UK.

It is impossible to name them all but these are the people who have helped me most. George Simms, sadly no longer with us, was the first to encourage and support me and a few of my colleagues, as we learnt the theory behind bees and beekeeping.

John Hendrie, an ex-trustee of the BBKA and colleague helped me on the road to teaching beekeeping and worked with me to prepare the first tutor pack in support of beekeepers running introductory courses. Ken Basterfield, the NDB education director, helped me to be able to 'read' the bees and learn from them rather than rely simply on what the books say. Everything I learnt from him has helped me to become a better beekeeper. Norman Carreck and David Aston have helped me along the way by always being willing to engage in esoteric discussions about the many aspects of bees and beekeeping and always being capable of answering any question I may have.

I would like to thank the International Bee Research Association (IBRA), for permission to reproduce images from Dr Eva Crane's book 'The World History of Beekeeping and Honey Hunting'. The anatomical illustrations are from Frank R Cheshire's book 'Bees and Bee-keeping vol.1. – Scientific'.

Finally, I would like to thank my wife Jan and close family and friends who have been forever patient and supportive of my interest. Jan has also worked very hard for beekeeping organisations at local and national levels and is well known and respected in beekeeping circles.

Dr Ivor Davis NDB
April 2015

Foreword

It is with great pleasure that I write this foreword to the second edition of the BBKA Guide to Beekeeping.

Taking up a new craft is both exciting and challenging and never more so than with beekeeping. We have co-existed with honey bees for a very long time and we increasingly recognise the importance that honey bees have played in providing sweetness to our diet, bees wax for light and most importantly the pollination of fruit and crops vital to human survival.

Honey bee dependence on us has been less obvious but it is the case today that honey bee health and survival is dependent on the beekeeper, especially because of threats from pests and diseases. So beekeeping is important and in order to do it well and benefit their bees the beekeeper needs to develop a range of beekeeping husbandry skills which will anticipate and meet the requirements of their bees.

Honey bees are kept in many differing kinds of environments in the UK and will exploit the sources of pollen, nectar and propolis available to them. Agricultural and horticultural crops also vary across the country and so healthy honey bees will have adapted to the local conditions. Beekeepers need to be aware of these variations and so it is no surprise that books on beekeeping contain a variety of beekeeping techniques and opinions on the best types of equipment to use and husbandry practice to follow. All of which can be confusing to the new beekeeper. Understanding the degree of commitment and time necessary, the equipment needed, sources and strains of honey bees available are also aspects on which guidance is essential.

This second edition contains some new text and updates the first edition which has been well received and I am sure both new and practicing beekeepers will find it a source of both information and inspiration.

I hope you will enjoy using the book and that it will be a constant source of reference for you.

Dr David Aston
President – BBKA, April 2015

Contents

1

Keeping bees has become very popular in recent years with the number of beekeepers in the UK increasing rapidly over the last five years. In 2007 the British Beekeepers' Association had 13,000 members; by 2014 this had increased to nearly 25,000. Beekeeping suppliers have all seen a rapid increase in the sales of equipment and all this has happened alongside press stories about the plight of the honey bee with major losses occurring across the world.

Almost a full load of pollen, being collected from a Ceanothus (*Ceanothus cordulatus*) before returning to the hive.

Main picture: Sealed honey stores surrounding cells of pollen, maturing larvae and sealed workers awaiting emergence.

The essential insect

It has been suggested that Einstein predicted that if all the bees in the world disappeared then mankind would follow in four years. This quote is wrongly attributed to him but life would be much poorer without the vast army of pollinators of which honey bees are crucial members.

We now realise that all bees are vitally important to the environment. If humans do not care for them by conserving their habitats and protecting them from the effects of modern living, we will suffer in the long term.

Worldwide the honey bee is extremely important as a pollinator of many crops. In California for example, the honey bee is critical to the production of almonds and over recent years, there has been concern that there may simply not be enough honey bees in the USA to pollinate the crop. If the almond blossom is not adequately pollinated the tree will be unable to produce almonds!

In the UK it has been estimated that all insect pollinators contribute more than £400 million to the agricultural economy at farm gate prices. This probably represents in excess of £1.5 billion, once the food reaches supermarkets. Honey bees are a significant contributor to this figure and it has been suggested that this one species could contribute up to 50 per cent of the pollination value. In fact, it is fair to assume that most flowering plants need a pollinator in order to propagate. There are some exceptions, but life would not be as pleasant – or as productive – without the vast army of insects that pollinate our flowers.

Most species of bees are in decline and the most effective way to reverse this trend is to improve their habitat, in large part by not practising agriculture in such an intensive way that all wild flowers and hedgerows are destroyed. These are the countryside features that favour the pollination force. Urban areas tend to be very good sources of food for bees because the variety of plants in gardens and parks provide an extended season for bees to collect food and provide a varied diet.

Honey bees are just one species in a vast range of bees. Most people know that bees produce honey, live in colonies and are pretty furry animals with a sting in the tail. In fact there are about 250 species of bee living in the UK. There are about 225 solitary bees that do not have a colony but tend a few larvae each year to produce a new generation and then die. Together they provide a major pollination force and, although they are

Colony losses

Colony losses have devastated bees and beekeepers around the world in recent years, a phenomenon that many people have tied to the use of insecticides widely used in agriculture. However, records show that over many decades there have been similar incidences of major losses and it cannot be assumed that current agricultural practices are the only reason for large losses. Indeed, bee populations vary from year to year and are closely related to weather.

Some European and US scientists suggest that losses of biodiversity and food resources, due to climate change, have intensified the problem.

Others believe that a rise in single-crop farming and the modification of landscapes, as well as pathogens causing diseases like foulbrood, Nosemosis and Varroosis are responsible for the problem.

Hedgerow flowers – here Red Campion (*Silene dioica*), require the services of pollinators in order to flourish.

Colony Collapse Disorder

While so called Colony Collapse Disorder (CCD) appears to have multiple interacting causes, some evidence points to pathogens and sub-lethal pesticide exposures as important contributing factors.

Laboratory studies show that some insecticides and fungicides can act together to be many times more toxic to bees than they would be if used alone. They can also affect the sense of direction, memory and brain metabolism, while herbicides may reduce the availability of plants bees need for food.

Agricultural and horticultural chemicals need to be used with care but should not always be blamed for bee losses.

wild and not tended or managed by us, are very important to the agricultural economy and the environment. In addition there are about 24 species of bumblebees; these are our furry friends that are seen working hard collecting pollen and nectar from a wide range of plants. Most are wild and survive by producing small nests each year. Whilst they collect nectar and store it in the nest the amount is very small and not harvested by us. However they provide an important pollination force and are bred to supply small colonies that remain in glasshouses to pollinate crops, mainly tomatoes. In these circumstances honey bees would not survive and be capable of providing an effective pollination force.

Then there is just one species of honey bee in this country and throughout Europe, America and Australasia (*Apis mellifera*). There are other species of honey bee living in Asia but this book will not deal with them in any detail. The honey bee is different from other bees in that it lives in a large colony and has strategies to survive through winter and periods when there is no food by storing excess when times are good. Whilst no one bee lives for long the colony can survive for many years. They are a bit like the old adage about the axe that has been in the family for many years and has had a number of heads and handles. Given the right conditions the colony can replace every component over time and continue to thrive.

Honey bees and humans

The honey bee is a special case. Like all other bees it is a wild animal, but humans have found it an advantage to work with colonies of bees. Although beekeeping has been practised for many thousands of years worldwide, recently honey bees have suffered major losses of numbers of colonies. New diseases have devastated colonies while modern farming practices may have exacerbated the situation.

Of particular note is the Varroa mite (*Varroa destructor*). This pest has infested virtually all honey bee colonies across the world. Untreated colonies seem to succumb to the mite and die out within two to three years. There is no simple solution to the

Honey bee collecting pollen from Greengage (*Prunus*) blossom in the authors' garden.

problem and, whilst some strains of honey bee are more resistant to the Varroa mite, it continues to be a major cause of colony death. The mite has also survived human attempts to eradicate it. It may be that in a couple of hundred years the honey bee will be able to coexist with this new parasite but in the meantime we need to help the bees to survive. The effect of the mite has been a dramatic decline in naturally occurring (feral) colonies and in general all honey bee colonies are now managed by beekeepers. Today's honey bee needs more humans to learn how to look after colonies and become beekeepers.

Keeping bees is an ancient tradition that is of great interest to young and old. If you go to a party and admit to being a beekeeper you will be inundated with questions about bees and may be in danger of blocking out all other conversation. Whenever beekeepers take a stall at a country show the public will be keen to buy honey and fascinated to see the bees in an observation hive where children will love to find the queen bee.

Beekeeping is essentially a practical skill that virtually anyone can learn given time. As with most practical skills, it is better to understand the reasons behind the various management techniques before actually starting to keep bees. Knowing why you are carrying out a particular technique will give you more pleasure and will also ensure that, with time, you can become a competent beekeeper.

What is a honey bee?

A honey bee is an insect; it has three components to its body: head, thorax and abdomen. It lives in a community known as a colony that exists through the winter and can continue to for many years.

A colony will normally have only one queen who will lay all the eggs that produce new bees. In the summer there may be 1000 to 2000 drones (male bees). Once autumn comes, the drones are ejected from the colony and will die. The majority of bees in the colony are workers (females) that carry out all the work required to maintain the colony. The queen is mainly an egg-

Colony size

Although the average size of a colony varies greatly – they can have between 10,000 in winter and 60,000 individual members, at the height of the summer – the proportions of different bees are as follows:

- *A queen*
- *1000 – 2000 male drones in summer*
- *10,000 – 60,000 female workers depending on the time of year.*

The greatest number of bees will be in summer when the queen is laying well and there is the maximum amount of pollen and nectar available for the colony.

The queen. Note the larger (longer) size of the queen by comparison to her attendant workers.

laying machine while the operation of the colony is controlled by the workers, acting not as individuals, but collectively. These little insects carry out the most amazing and complex tasks requiring close cooperation including swarming, or defending their colony from predators.

In the UK honey bees are unique in that they continue to live as a colony throughout winter when there is no food to be collected from the flowers. They have evolved to be able to store food in the form of honey so they can survive when there is no other source. This remarkable development to get through difficult times means that bees store honey and, under the right conditions, will store more than they need. This 'excess' provides us with the honey that has been so popular with humans – and many other animals – for thousands of years.

Solitary bees

Solitary bees, as the name suggests, are a group of species where the adults live as individuals. The females lay eggs and then provide each one with nectar and pollen in underground hollows or tubes. The eggs hatch into larvae, which develop by consuming the food, placed around them by the adult female that laid the egg. Fully-grown larvae then pupate and eventually emerge as adult bees. Male and females will mate and the process will start again in the following year. Depending on the species, winter is spent as either a fully-grown larva or adult. In collecting pollen for their larvae they pass pollen from one flower to another and thus pollinate them.

Top image: a Common Carder Bee (*Bombus pascuorum*) taking nectar from a Blue Spiraea (*Caryopteris*).

Lower image: a Buff-tailed Bumblebee (*Bombus terrestris*) collecting pollen from a bramble (*Rubus fruticosus*).

Bumblebees

Bumblebees live in a small community during spring and summer. In autumn, the colony will produce a number of queens and drones (male bees). These will mate and the fertilised queens will then spend the cold winter months sheltered in a protected place such as an uninhabited mouse hole or other relatively 'warm' place. In the spring the queens will emerge, feed and start a new nest where they will raise worker bumblebees that take over the duties in the nest. The nest will eventually break down in autumn after producing a new crop of queens and drones. Bumblebee nests are only used for one season.

Animal husbandry

Anyone who takes up beekeeping must remember that they are taking responsibility for caring for animals. Bees may be small and somewhat alien but they are still animals with needs and a beekeeper must ensure that bees are kept in the best of health and in conditions that are acceptable to them. It has been shown many times that bees may become difficult to handle if the conditions under which they are kept are poor. We will deal with all the aspects of good husbandry later in this book but it is important to consider the following factors before you start beekeeping:

- Are you ready for a fascinating new activity in your life?
- Are you prepared to learn properly how to manage bees?
- Will there be enough forage for the bees to prosper?
- Could the bees become a nuisance to neighbours and other members of the public?
- Are you prepared to spend time regularly attending to your bees?
- Will you ensure that the bees have sufficient honey stores and that you will not take more than they can afford to give you?
- Are you prepared to collect a swarm if required?
- Are you prepared to get stung on the odd occasion?
- Will you always wear clean and appropriate clothing when working with bees?
- Are you ready for friends and neighbours to ask endless questions about beekeeping?

At this early stage it will not be possible for you to answer all these questions so it is always recommended that you join a local beekeeping group who can help you to get started and teach you the basics of the craft.

There are over 300 groups in the UK who are there to help. They are usually locally based and run by amateurs (people who love bees and beekeeping). Even after you have been keeping bees for a few years something will happen to your bees that may be difficult to understand, such as a colony that is not thriving when it should be or a queen that is not doing what is expected of her.

Inspections

In winter you will only occasionally need to take a look at your hives to ensure that they are in good shape and not damaged.

In summer, more time is required and you will have to monitor your bees weekly when the colony is strong and nectar is in full flow. This may only take 15 to 20 minutes per hive as you look carefully through all the frames in the brood box checking for stores – honey and pollen – and any signs of pests or diseases.

A quick inspection of a full brood frame of bees. It will be necessary to shake off the bees over the brood box to check more carefully for any signs of disease.

Antennae

The main sense organs of the honey bee – the antennae – are very adaptable, so much so that bees have been trained to detect explosives by the USA military.

However their primary use is in sensing the pheromones of the queen and in communicating with other bees in their home colony.

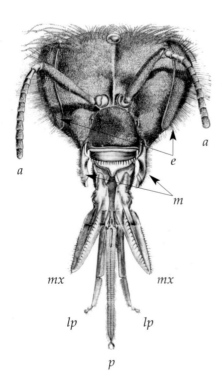

The head of a worker bee from an illustration produced in the late 1880s.
a, antennae (feelers); *e*, eyes;
m, mandibles (outer jaw);
mx, maxilla (inner jaw); *lp*, labial palps;
p, proboscis (tongue)

This is when a local group can be very helpful because you will always be able to call on another beekeeper that will have experienced your problem and can give advice. You will also learn that if you ask three different beekeepers about your problem you will get at least four answers! Probably all four answers are correct and you just need to find the approach that suits you and your bees.

As with all animal husbandry the bees do not read beekeeping books and will decide quite independently what they need to do when faced with a particular circumstance.

The anatomy of a honey bee

There are male bees (drones), and females bees (queens and workers). The following anatomical descriptions apply mainly to workers and include some information about the differences between workers and the queen and drones where appropriate.

The head

This includes the eyes, antennae, mouth, mandibles, the brain and a food production gland all neatly packaged into a hard casing that is about 3mm square and about 2mm deep.

The eyes are compound with each eye consisting of between 6500 to 9000 facets (depending on whether the bee is a queen, a worker or a drone). Each facet can receive light from a narrow field of view that is different from all the other facets. This gives the bee the ability to perceive the world somewhat like looking at a digital photographic image made up of only a few pixels.

The bee can perceive colour, although it cannot receive red light. Instead it is able to see ultraviolet light (light we cannot see). The compound eyes are also able to perceive polarised light in the same way as humans can when wearing polarised sunglasses. This gives the bee the ability to sense direction relative to the sun even when the sun cannot be seen! As the eyes cannot move with respect to the head, the bee must turn its head to see in another direction (similar to headlights on a car). The honey bee can see a wide field of view from each side and vertically. The male bee (drone) has larger eyes that meet in the middle and is able to see well straight ahead. In addition to the two compound eyes, the bee also has three light sensors on top of its head. It is believed that these allow the bee to accurately determine where 'up' is and help to maintain stability during flight. They may also be used in some way to control the more complex compound eyes.

Honey bees also have a pair of antennae. They are attached to the front of the head and can be moved by the bee. They are covered with an array of sensory organs that allow the bee to sense its environment. The primary sense is smell and there are about 30,000 sense organs on each antenna. It is thought that some are specific to the smell of just one chemical compound that play a part in bees' extremely sensitive perception of pheromones (chemical messengers that are used to communicate between the bees).

A bee's sense of smell is so sensitive that it is used to provide warnings of illegal drug transport or explosives. They are even being used to detect cancers in humans through smell! The antennae also provide the sense of taste and touch and there are other sensors on the antennae that allow the bee to measure wind speed.

Bees have a very good sense of smell and with two antennae they can perceive smell in 'stereo'. In effect, they are able to recognise small differences in the concentration of a smell from each antenna and so work out the origin of a smell.

The mouth is similar to a tube with a suction pump in the head. The proboscis – the tube through which the bee can draw up liquids – is normally folded away beneath the head but when the bee wishes to drink nectar or water the proboscis is brought forward and held by the mandibles like a straight drinking straw. On the end there is a brush so that if the food is not liquid the bee can drop saliva onto it to make a liquid by brushing before sucking it up into the head. The pump in the head is then able to push the food back into the stomach. Adult bees normally only consume nectar, honey, water and pollen. The first three are naturally liquid and the last is mixed with diluted

Life of a worker

It's a tough life and a short one for a worker bee.

Workers raise young, build the comb, take care of the queen, carry out guard duty at the hive entrance, remove any dead bees, provide heat when required and cool air when it's hot – they also go out and forage for nectar and pollen.

What makes worker bees all the more interesting is that their tasks within the hive change as they age.

Bees by nature are very clean insects and will deal quickly with any honey or syrup that a beekeeper accidentally spills. You can clearly see the tongue of the lower bee – the orange tube like proboscis – seen sucking up the liquid honey.

Protection

A strong colony of bees will protect both their stores and their larvae from marauding invaders.

Weaker colonies may suffer badly from robbing by other bees and invasion by wasps and hornets who are after larvae to feed to their own young.

Beekeepers need to ensure that with weaker colonies, suitable wasp or hornet traps are used at vulnerable times during the season if they are to preserve the colony.

When baiting wasp traps – ensure that you don't use your precious honey. A little jam or marmalade is preferable, and will help to contain the spread of any disease.

Details of hornet traps and how to deploy them are available from the BeeBase website – www.nationalbeeunit.com

honey before being ingested. Sometime bees will consume any sweet liquid such as honeydew from aphids or even molasses spread on fields by farmers.

Bees have a very narrow waist between the thorax and the abdomen (where the stomach is located). For this reason a bee is only able to ingest liquids (solids would get stuck in the oesophagus). The mouthparts are therefore designed for collecting fluids only.

The mandibles are two hard shovel-like plates on each side of the mouth. When the proboscis is retracted, the mandibles can be used to 'chew' any appropriate material. They are frequently used to shape pieces of wax when building honeycomb.

When bees fight they sometimes use their mandibles to cut the legs off their opponents or tear pieces off their wings. They also use their mandibles to chew and form propolis (an antiseptic sticky substance collected from the buds of flowers) which is used to fill holes in the nest and cover the surface of honeycomb cells prior to the queen laying an egg in a cell.

The brain is well protected inside the head and comprises two lobes that process the information that comes from the eyes, antennae and other sensory cells. As with human brains there is an area that will analyse these inputs and apply logic to the subsequent actions. In the honey bee this is done by a few hundred cells – a remarkable feat of miniaturisation.

Unlike humans the honey bee has ganglia (groups of nervous cells) strung along the body. These cells provide autonomous action for local parts of the bees and thus reduce the load on the brain.

The food production gland is called the hypopharyngeal gland and is located in the head. The gland is very important as it produces some of the food essential to raise new bees. This very rich, high protein food is mixed with a substance produced from the mandibular glands. These are the essential ingredients of royal jelly, the food fed to larvae that are intended to become queen honey bees. As a worker bee ages, it will complete its duties in the hive, feeding larvae and protecting the hive to become a forager (a bee collecting pollen and nectar from the environment). At this stage the hypopharyngeal gland will evolve to produce enzymes that are essential in the process of making hone from nectar.

The head, like the rest of the outer surface of the honey bee is made up of a number of plates – formed of chitin, a material similar to human nails (keratin) – that are connected by softer membranes. This means that the shape of the head can be varied slightly but also that the hard outer structure provides a framework for 'suspending' internal organs and obviates the need for a skeleton.

The head is connected to the thorax by a flexible tube, also made of chitin. Muscles connecting the head to the thorax allow some movement of the head relative to the thorax and the bee can move its head to be able to 'see' in different directions.

The thorax

This is the second part of the honey bee's body and can be likened to the powerhouse of the insect. It comprises wings, legs and powerful muscles used in locomotion and flight and to generate heat.

There are two pairs of wings: a fore pair and a hind pair, one of each on either side of the bee. The wings are very fine membranes with stiffened veins holding the profile. The wings can be folded backwards so that when the bee is not flying they lie neatly over the body and do not get in the way. The forewings are constructed so that when the thorax is compressed in a vertical plane they move upwards and when the compression is released the wings move down again. This means that the power needed for flight is generated by the bee changing the shape of its thorax. Very small muscles between the outer body of the thorax and the wings allow the bee to change the shape of the wing and allow it to fly in any direction. The hind wing is not powered, but when the forewing moves out across the body in preparation for flying, hooks on the rear of the hind wing

Top: Connection between the head and thorax allows limited movement of the bee's head.

Bottom: A worker with pollen grains on her head and thorax going over an Abelia (Abelia x grandiflora).

Temperature

Worker bees will control the temperature and humidity of the hive throughout the year.

In winter they will huddle together, and as happens with penguins the outer layer of bees in the cluster will move towards the centre as they cool down.

In summer when the hive is too warm the bees will fan their wings to induce a greater flow of air to cool down the colony – rather like our air conditioning systems.

The real killer for a colony of honey bees is dampness and not as you might expect severe cold.

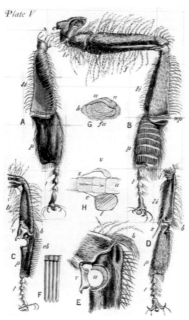

Plate V

Legs of a worker bee – A and B show the hind leg, C a front leg and D a middle leg.

connect into a groove on the fore wing and fix them together, giving a large surface area for improved flight.

Honey bees, like all insects, have three pairs of legs. They all originate from the thorax. Apart from allowing the bee to walk they have specialist features that the bee can use during its life. The most specialised legs are those of the worker bees. The fore legs have a notch, located in one of the joints, that is used to clean the antennae. When the bee is ready to fly it will wrap its antenna cleaner around the base of the antenna and draw the antenna through the notch. This removes any debris and ensures all the sensors are able to operate efficiently.

All the legs are hairy and, when the bee is hovering, they are used to collect pollen off the rest of the body. The pollen is mixed with nectar from the bee's mouth and she then uses her legs to pass the sticky pollen backwards to the rear legs. The rear legs are very specialised and have upward-facing combs on the inside of their lower part. These move the sticky pollen upwards when rubbed together through the adjacent joint, where the pollen ball is compressed and pushed outwards into a pollen basket on the outside of the leg.

The basket is made from long hairs that curve out from the leg to produce a case for the compressed pollen. The worker bee's legs are very well adapted to permit the bee to collect pollen from its body and store it on the sides of their back legs.

Probably the most remarkable organs in the thorax are the **flight muscles**. These are a pair of very powerful muscles that deflect the shape of the thorax and drive the wings. In so doing they produce much heat which is dissipated through the body of the bee by the flow of haemolymph (bee blood) through the thorax which raises the overall temperature.

When the wings are laid over the back of the body, the wing muscles are still able to flex without moving the wings. In this state the bee can use the muscles to generate heat. The heat is important because the brood (eggs, larvae and pupae) need to be kept warm in order to develop properly.

The bees keep the brood at a constant temperature between 33°C and 34°C irrespective of the external temperature around the hive. In winter, when the external temperature can be very low, the bees will cluster together and use their wing muscles to generate heat to ensure that none of the other bees' internal temperature drops below about 8°C. Below this temperature honey bees become torpid and will soon die.

The abdomen

The abdomen is the back end of the honey bee. It is connected by a narrow 'waist' (called the petiole and also made of chitin) to the thorax. The abdomen holds a number of key organs and functions the bee needs to survive such as the digestive tract, the heart, wax glands, the Nasonov gland and the sting.

The bee has a stomach (known as the ventriculus) where it digests pollen and nectar. In front of the ventriculus there is a sac called a honey stomach that can expand to hold nectar gathered from the flowers. It has a valve (the proventriculus) connecting it to the ventriculus that filters out pollen from the nectar and passes the pollen with some nectar back into the ventriculus. The nectar can then be regurgitated when the bee returns to the hive and passed on to another bee for storage.

At the back end of the ventriculus there is another valve (the pyloric valve) that controls the passage of food from the

Bees and smoke

When a hive detects smoke, many bees become remarkably calm; it is speculated that this is a defence mechanism.

Wild colonies generally live in hollow trees and when bees sense smoke it was assumed that they prepared to evacuate their nest after gorging on honey stores.

However it is now thought that smoke stops effective communication and the result is that bees cannot mount a defensive attack.

These two images clearly show the large size of the female worker's abdomen.

Left: a worker looking for nectar on a Blue Spiraea (*Caryopteris*) flower.

Above: worker bees checking their honey stores.

Beeswax

Workers will need to consume around 5 litres of honey to produce enough wax for a Standard National frame – see page 41 – of honeycomb.

When bees build new comb onto the wax foundation, this is known as 'drawing out' the comb.

Wax is squeezed out of the bees' wax glands as a clear liquid which solidifies quickly and turns white.

Workers use this wax to produce the beautiful comb that we see here on the right.

ventriculus to the intestine, very much like in humans. The digested food is absorbed by the intestines and the waste matter passed to the rectum at the rear of the abdomen. The rectum can expand enormously and hold a lot of waste matter. This helps in winter when inclement weather can mean that a bee is unable to leave the hive and evacuate any waste matter. Bees are extremely hygienic and never normally defecate in their hive.

Worker bees produce the wax used for building honeycomb and joining parts of the nest together. To do this they cluster together to raise the body temperature to about 37°C. Each of the workers has four pairs of glands on the underside of the abdomen that produce small thin flakes of wax. The bee collects the wax with its legs and passes it to the mandibles where they chew it to make it 'workable'. They can then fashion the wax as required. Usually this is in the shape of hexagons that neatly fit together.

Above: Wild comb built by workers into the roof of a colony that has clearly not been looked after by a beekeeper.

Right: Polished cells, some with eggs, showing the symetry and inbuilt skills of worker bees in building comb ready for the queen to lay.

It is believed that when working the wax, the bee forms a tube around itself and that this process helps to heat the wax slightly. When this happens the connecting strips of wax between each 'tube' tighten to produce the straight sides of a hexagon. The same process can be seen with soap bubbles on a flat surface whenever there is a group of bubbles all about the same size. If you look carefully you will note that the bubbles are not circular but are hexagons that fit neatly together.

The Nasonov gland is a specialist gland on the top of the abdomen near the rear end. The bee can expose this gland by bending down the last segment of the abdomen. The gland emits a pheromone (a scent that influences the behaviour of another bee) that attracts other bees to the individual. In this way the bees can collect together whenever they are disturbed and flying bees can leave a scent trail to encourage other bees to follow it.

The sting is an organ situated at the back of the abdomen of a worker. It is an evolution of the ovipositor used by the ancestors of bees to lay an egg in a cavity or another animal. This is the reason only females have a sting. Queens have a sting but this is only used against another queen. Should there be two queens in a colony it is usual for them to have a fight and once one has overwhelmed the other it will sting the defeated queen behind the head to kill her.

The worker's sting is used for the defence of the colony. It is unusual for a worker to sting in any other situation except when it panics or becomes trapped (in your hair for instance). When the worker stings a large mammal like a human it will normally die.

The sting is barbed and locks into the skin. If the bee tries to fly away the sting and the nerves and muscles operating it are torn away and she will die a few hours later. The advantage of this is that the sting will operate without the bee. The sting shaft is made up of two semicircular canals that form a tube through which the venom is pumped into the victim. Each of the canals has barbs on the outside and alternatively moves up and down. The action pushes the sting further into the body. So, once the bee has gone, the sting continues to penetrate the skin and pump venom into the victim. It is even equipped with a pheromone gland that attracts other worker bees to sting in the same place. This mechanism is so effective that bees are able to establish a mass attack and drive away almost any animal that attacks the colony.

Drones do not have a sting and are not involved in protecting the colony.

A sting in the tail

Bees are not by nature angry or aggressive; however they are defensive. If threatened, handled poorly or their colony is attacked by farm animals or pets they may respond by stinging.

Bees' lancets are barbed so it is difficult to remove them from the victim. When making her escape the worker's sting mechanism is torn away from her body and she dies slowly.

Top: Bee preparing to sting. Note the arched abdomen.

Below: Quickly scrape off the sting sack and sting with a thumb or hive tool.

■ *The colony*

2

Honey bee colonies are often referred to as 'superorganisms'. This is because each individual honey bee is unable to live alone or do much to support itself, but as a group they are able to organise themselves to behave as a single, highly effective organism.

Right: Busy workers bees attending to their stored honey.

Above: a worker inspecting an Abelia flower (*Abelia x grandiflora*) for pollen. If you look closely you will notice her middle legs are in the act of brushing pollen from her thorax.

- Adult bees leave the hive to find and bring back food.
- They collect for all the needs of the hive and not just their own requirements.
- Inside the hive the young bees move around and collect debris or dead bees to keep the hive clean and free of inherent disease.
- Guard bees will stop invaders entering the hive and robbing its content.
- Young bees and the queen produce and feed the young larvae to create new bees.
- The comb in the brood nest accumulates waste matter and locks away diseases.
- The communication in the hive through pheromones and food sharing allows the bees to work together for the benefit of the hive. Adult honey bees will die if they use their stings yet they do sting intruders for the benefit of the hive.

Honey bees, ants, termites and other animals that live in communities demonstrate these characteristics. What is interesting is that invariably these communities are dominated by females that have given up the role of procreation to a 'queen' and will work together for the benefit of the community.

Individual honey bees are unable live on their own and need to be part of a colony in order to survive. If a honey bee becomes isolated from its parent colony it will try to join another colony by begging entrance to the hive. If it does not find another colony it will die within a few days.

For a colony to survive it must have four components:

- A queen
- Worker bees
- Drones (during the summer months)
- A nest with brood (eggs, larvae and pupae)

If one of these components is missing the colony can become stressed, out of balance and therefore prone to disease and less productive. The workers will try to re-establish the balance as soon as possible by:

- Making a new queen
- Encouraging the queen to lay more eggs
- Encouraging the queen to lay drone eggs
- Creating a new nest or repairing the existing nest

The brood nest is the heart of a colony. Here, new bees are nurtured as eggs and larvae that eventually pupate into adult bees. Unlike many other insects, the larvae in a bees' nest remain in cells and are fed by adult bees. The temperature is raised so that the development time can be shortened. The centre of a bees' nest is kept at a constant 33° to 34°C and with a humidity of about 40 per cent. These are ideal conditions for larval growth.

The queen bee

There is normally only one queen in the colony. She is female and under ideal conditions can live up to five years. She has the same-sized head and thorax as a worker but her abdomen is al-most twice as long and her legs too are longer. The queen's main job is to lay eggs that will develop into new bees. Worker bees

The queen bee

Once the queen has mated successfully she will remain in the colony for the rest of her life (up to five years).

The normal life cycle of a very productive queen is usually two to three years, after which time the bees or beekeeper will replace her.

She can store up to eight million sperm in her body and this is enough to keep the colony operating for many years.

An unmarked queen surrounded by her attendant workers.

Laying eggs

A good queen – a well mated, young queen – will lay upwards of 2000 eggs a day at the height of summer.

Here we see a brood comb (right), along with food stores – liquid honey and pollen – larvae (grubs) in various stages of development and sealed worker cells.

The picture shows 'patchy' laying – this may be due to a young queen or other causes.

The observant beekeeper will look out for disease amongst larvae in the unsealed cells.

feed the queen with brood food. This is a mixture of secretions from the worker mandibular and hypopharyngeal glands supplemented with some honey. Brood food produced in glands in the head of worker bees is high in energy and nutritional value. The more the queen is fed, the more eggs she will lay.

In spring and summer when the colony is expanding rapidly the queen may lay up to 2000 eggs in a day. The weight of that vast number of eggs is greater than her bodyweight. She does not really have enough time to feed herself or to digest pollen and nectar (the normal food for adult bees). The food fed to the queen by workers has been digested and the proteins have been converted into essential amino acids that the queen can use directly to produce eggs. As the season moves on, the number of eggs laid by the queen diminishes, and in the depth of winter the queen may cease egg laying altogether. As the need for new bees is reduced, the workers cut down the food they give to the queen to reduce her egg-laying capacity.

The queen produces pheromones that control the activities of the workers. One effect of these is to prevent the workers (who are also female) from laying eggs. Another is to let the colony know that there is a queen in the hive. If the queen is removed for any reason the colony will recognise this within 30 minutes and start preparations to produce a new queen.

Cells full of eggs in the image above show that the queen has been hard at work laying to ensure the colony's viability.

The pheromones produced by the queen are spread throughout the colony by the workers touching her and then touching other workers. Often, when you are looking out for a queen in a colony, she can be seen surrounded by a 'court' of workers who are feeding her, touching her and gently guiding her round the nest to encourage her to lay eggs.

There can be more than one queen in the hive. There is a process called supersedure when a mother and her young daughter will both happily lay eggs as they move around the colony. On other occasions there may be two or more queens raised by the colony but if this occurs the queens will find each other and engage in a fight where one will die. The remaining queen is unhurt and will proceed to lay eggs for the colony.

Royal jelly

Royal jelly is a creamy white substance, responsible for giving the queen a long and healthy life.

Royal jelly can kill many strains of bacteria and when fresh is a powerful disinfectant.

Royal jelly is a useful source of vitamin B5 and is used in some cosmetics for its anti-ageing effects.

The development of the queen

A queen is produced from a normal fertilised egg. Once the egg hatches, the workers will feed the larva with an exceptionally rich food, high in sugar and nutrients known as royal jelly. The larva is provisioned with more food than it can consume so is never lacking food. The outcome is that the larva grows extremely rapidly and when fully grown the body contains high quantities of juvenile hormone. The larva grows to about 2500 times the weight of the egg within just five days. When the larva pupates the juvenile hormone results in the body forming into a queen. It is believed that the hormone causes different genes in the DNA to be switched to produce a queen instead of a worker bee. In effect the DNA of a female honey bee describes two different insects and the way the genes ultimately express themselves is controlled by the way a larva is fed.

A nice queen cell ready for cutting out and using in a colony that needs to be requeened.

Workers are developed in normal wax cells but queens are produced in special queen cells that hang vertically down (as opposed to lying nearly horizontal). Under normal conditions these cells start as 'queen cups' that look a bit like an acorn cup. Once the egg has been laid in the cup the workers start to fashion the sides of the cell, which grows as the larva develops. Once the larva is fully-grown the bottom of the cell is sealed with a porous capping made from wax mixed with pollen. The larva pupates inside the cell for a further eight days and then is ready

Mating

Drone collection areas are often nowhere near the hive, so it's a minor miracle that queens are mated at all.

The mating areas are always well above ground and rarely seen by beekeepers, as is the actual act of mating.

to emerge as a fully formed queen. Workers are able to control the timing of a queen's emergence because the wax around the cell is too thick for the queen to chew her way out of the cell. They will thin the base of the cell when they are ready to release the new queen.

The act of pupation (forming the adult body from a larva) uses all the energy available to the queen. Once she emerges from the cell she will rush off and feed. It takes about four days for her to be able to produce queen pheromones and during this time she is left on her own. Then the workers become aware of her existence, start to feed her and encourage her to leave the colony on mating flights.

Above: A drone collection area with a queen attached to a fishing pole for the photograph.

Right: Newly laid eggs in these cells, where you can in some cases also see eggs laid in the cells behind those being photographed.

Mating and reproduction

The virgin queen flies from the colony with a small entourage of worker bees. She goes to a specific area near the hive where drone (male) honey bees gather in what is known as a drone collection area. Here, the drones will chase the queen and the strongest and fastest-flying ones will catch her and mate with her 'on the wing'. She will mate with up to 20 drones and store their semen within a special gland (called a spermatheca) in her body. However, she can only be mated within about three weeks from achieving maturity. If unsuitable weather or another obstacle stops her from mating in this period she becomes sterile and will only be able to lay unfertilised eggs (which produce drones). From the colony point of view she is rendered useless and usually the whole colony will eventually die.

Once the queen has mated successfully she may remain in the colony for the rest of her life (up to five years). She can store up to eight million sperm in her body and this is enough to keep the colony operating for many years.

Once she **starts laying eggs** the queen will remain in the nest and normally only leave when the colony decides to swarm. When this happens, the queen will leave with about half of the workers and find a new nest site. The old colony will create a new queen and continue to operate. Swarming is a natural process of colony reproduction.

Drones

Drones are male bees. Their primary job is to mate with queens. They are accepted into any colony and seem to move fairly freely about in an area and visit any number of colonies. Drones develop from unfertilised eggs laid by the queen.

The workers build the honeycomb cells in the colony. When they do this they are able to determine the size of the cell. Workers' breeding cells are slightly smaller than those used for breeding drones. When a queen lays an egg she will feel the size of the cell with her front legs and antenna. Once she has sensed the size she will reverse into the cell and lay an egg at the base of the cell. If the cell is a large (drone) cell she will leave an unfertilised egg. If it is a smaller (worker) cell she will deposit a few sperm onto the egg just before she lays it. The sperm penetrate the egg lining and fertilise the egg.

In this way the workers are able to regulate the ratio of workers to drones in the colony by guiding the queen to lay eggs in cells for workers or drones.

Drones are larger than workers and take longer to develop. During the first two days after the egg hatches the workers will

Drones

Drones do not keep house!

They cannot produce wax, do not forage, clean out the combs or guard the hive entrance.

In short they are pretty expensive for a colony to support.

In the late summer, the queen will cease to lay any more drone eggs and as forage becomes more scarce the drones are finally expelled from the hive and will perish without its support.

Look at the size of a drone in comparison to the worker bees! Their eyes are much larger and almost twice the size of those of the workers.

feed the larva with a mixture of secretions from their mandibular and hypopharyngeal glands similar to the feed given to workers. After the two days the workers will start to feed the larva with pollen, honey and a much-reduced feed from their glands. The larva takes seven days to grow to full size and then pupates in its cell for a further 14 days after the workers have capped the cell with a porous mixture of wax and pollen.

Development and lifespan

When the drone emerges he is fed for a few days with brood food by the worker bees. After that he will then learn to feed himself with pollen and nectar in the nest, but will still beg food from the workers in the hive. He does not fly out to forage for food. After about 10 days he will be sexually mature and able to leave the colony to go out on mating flights in search of virgin queens.

Drones live for about 40 days although this figure will vary with the weather and their level of activity. By the time September comes, all the drones will have been forcibly ejected from the colony by the workers. This is because after this time it is unlikely that the drones will be needed to mate with queens and as food will be scarce in the coming months the colony decides the drones must be sacrificed. Once a drone has been ejected from a hive it will only live for about two days before dying of starvation. The remainder of the colony endures through winter and early spring with no drones.

In late spring in the following year the workers will begin to guide the queen towards the larger drone cells and she will start to lay unfertilised eggs to raise the new season's drones.

At the height of summer there may be about **200 drones** in a colony although there can be as many as 1000. It appears that when a colony is preparing to swarm there will be more drones produced; in addition, other drones will be attracted to the colony in readiness to mate with any new queen. A colony with a number of drones in it often seems calmer. It has been suggested that, because drones are larger than workers, their presence in the colony helps to regulate the temperature in the nest.

Drone cells are larger and have a domed cap. Here are a group of capped drone cells that are surrounded by worker cells and some larvae that are maturing and will shortly be capped over with a mixture of wax and propolis.

Adding a small amount of propolis gives strength and the characteristic colour to cells that contain larvae.

Honey stores capped with pure wax are almost white in appearance and not porous. Pollen stores remain uncapped.

Workers

Workers are sterile female bees. They are produced from a fertilised egg laid by the queen. When the egg hatches, workers will feed the larva a nutritious food from their glands. This food is not as high in sugar or as nutritious as the feed given to queens. After two days the workers will start to give honey and pollen to the larvae. Feeding is done regularly but the larva is not given too much food (as is the case with the queen larva). The result is that the larva takes six days to grow and is not as large as the queen or drone larva. However it is still about 1500 times larger than the original egg when fully grown.

Once the larva has reached its full size, the cell is sealed with a capping of wax and pollen. The larva pupates and remains in the cell for a further 12 days after which the worker cuts her way through the capping to emerge as an adult.

In any colony, **workers are the most numerous members** (up to 60,000 individuals in the summer and down to about 15,000 in winter). The workers do all the work to maintain the colony and live for only about six weeks in summer; in effect they die from hard work. In winter, workers can live up to six months, keeping the queen and themselves warm and being ready to care for new brood once the spring returns and there are plants providing nectar and pollen again.

Egg laying workers

Workers are able to lay eggs and whilst there is a queen and brood in the colony, this ability is suppressed.

However, without a queen and brood being present, some workers will start to lay eggs.

An emerging worker bee shown by the arrow. She has successfully cut off the cell cap with her mandibles and now faces the long struggle to work her way out of the cell which may take several minutes to complete.

The picture (left) shows workers busy checking stores on some sealed and un-sealed honey comb.

Who does what in the colony?

When a new worker bee emerges she is unable to do many of the tasks in the colony. Her first job is to feed to restore her energy levels. She then starts work cleaning the hive by removing any debris that may have accumulated. As she grows older she is able to do other tasks and in a rough chronological order she will do the following duties:

- Clean the hive
- Clean and polish cells
- Feed larvae
- Feed the queen
- Make wax, build honeycomb and seal cells
- Process and store nectar and honey
- Guard the colony entrance

and then:

- Forage for nectar and pollen
- Forage for water and propolis.

During the active season, the first group of tasks is performed in the first three weeks of a worker bee's life and the last two tasks are done during the last three weeks of its life. This has the advantage that bees which have reached the end of their lives normally die whilst on a foraging trip and not inside the hive. Unlike modern humans the worker bee is expected to work to the end of its life and there is no concept of 'retirement'.

The change of the tasks **as the worker ages** is closely related to the development of the glands of the bee over time. The emergent worker has spent all her energies in pupating and must feed heavily on pollen and nectar so that her body can continue to develop. The first glands that start to work are the hypopharyngeal glands in the head of the worker; these produce brood food for the larvae and also food for the queen.

Top: Workers on guard duty at the hive entrance.

Bottom: Workers foraging on Helenium (*Helenium autumnale*) in late summer

Later, glands on the underside of the abdomen start to produce wax so that she is able to build cells and repair the nest. Later still the hypopharyngeal gland evolves further to produce the enzymes needed for the processing of nectar into honey. Finally the venom and sting glands become functional and the worker is then able to take on guard duties. The whole process takes about three weeks in summer and during this time the worker will remain in the hive for most of the time. Once the worker is fully functional she is ready to leave the hive, cease being a 'house' bee and become a foraging bee.

Genetic diversity is maintained in the hive as the queen mates with many drones and some workers appear to be more adept at some tasks than others. Also not all of the workers carry out all the tasks and, in an emergency, workers can revert and take on the duties of their younger sisters if required. Although their roles may seem very clearly defined, in fact worker bees are able to take on many different tasks during their short lives.

wg

The two rows of four wax glands are shown on the underside of a worker bee.

It is popularly believed that workers spend all their lives working for the good of the colony. Some detailed research has shown that a worker may in fact spend up to eight hours a day resting (or at least not doing any particular activity). Whilst the main time for activity is during the day when bees are foraging for provisions to take back to the hive, bees will continue to work in the hive overnight.

The brood still requires feeding; the nest must be kept warm and nectar still needs processing into honey.

Above: the many colours and sizes of bees working in a colony.

Lifecycle Table

Stage	Queen	Worker	Drone
Egg	3 days	3 days	3 days
Larva	5 days	6 days	7 days
Pupa	8 days	12 days	14 days
Emerge as adult*	16 days	21 days	24 days
Fully developed**	20 days	42 days	34 days
Lifespan (as an adult)***	2 to 5 years	6 weeks (summer)	3 months
		6 months (winter)	

*Emerge as adult: the time between an egg being laid and the adult emerging from the pupa.

**Fully developed: the time between an egg being laid and the adult being sufficiently mature to take on full duties in the hive.

***Lifespan: the expected age of an adult bee before death.

A new queen

At some stage in the queens' life, her pheromone levels reduce and the workers start to prepare for her replacement.

Once this happens, the bee-keeper will see some queen cells being prepared and will know that the colony is getting ready to replace the queen.

This process can happen anytime from early spring, but is more often seen in the late summer when the old queen is failing to lay sufficient eggs for the colony's survival.

Honey bee communication

Honey bees have **many ways of communicating** with each other. This is necessary because the safety and viability of the colony depends on the whole community working together. Most of the colony management is done collectively by the workers that, through a number of simple mechanisms, can recruit each other to perform tasks.

Most of the communication is done **using pheromones** (chemical messengers produced by one or a group of bees that influence the behaviour of other bees) but they also perform dances and share food to signal the location of a specific source of food. Dances are also performed to give fellow workers information, such as the time to swarm. These mechanisms have developed to be very effective and ensure the stability of the colony.

The importance of pheromones

The queen plays a strong **role in maintaining stability** in the hive. She produces pheromones that let workers know she is present. In addition her pheromones prevent the workers' ovaries from developing fully and thus inhibit them from laying eggs. Even the brood (larvae and pupae) produces pheromones that help maintain the stability of the colony.

Pheromone	Produced by	Effect
QMP (Queen mandibular pheromone)	Queen	Queen is present do not try to make another queen; Prevents worker ovaries developing
Brood pheromone	Brood	Stops worker ovaries developing
Nasonov	Workers (Nasonov gland)	Join me or follow me call from one worker to others
Alarm	Workers (mandibles)	Help me defend the colony
Attack	Workers (sting)	Sting this object to deter an attack on the colony

The pheromones work as a sophisticated control system that ensures that while there is a queen and brood the workers will get on with their normal duties of caring for the colony and providing food. The pheromone control system further ensures that there is a balance between young 'house' workers and older 'forager' workers. Research has shown that if all the forager bees are removed from a colony then the house bees will develop much faster in order to become foragers. Similarly, if all the house bees are removed, then some of the foragers will regress and become house bees.

The table on page 26 lists some of the more prominent pheromones produced by bees in a colony but there are many more. Most pheromones are a mixture of volatile and oily chemicals. The volatile ones have an immediate effect on the colony whereas oily ones seem to have a long-term effect. Our understanding of pheromones and their detailed chemical make-up is not complete. It is clear that bees have very sensitive 'smell' and 'taste' organs on their antennae (more sensitive than a dog's sense of smell) and each pheromone may have a number of different effects depending on the circumstances in which it is deployed.

One important effect is that the specific mix of pheromones, combined with the state of the hive and the food that has been collected, gives each colony a distinctive odour that is transferred to the occupants. This 'hive odour' allows bees to recognise their own colony and guard bees to recognise their colleagues. It is just part of the range of information that guard bees use to recognise intruders that may be trying to rob honey or attack the colony.

Food sharing

Honey bees spend a lot of time exchanging nectar. When a bee returns from a foraging trip she will pass her load of nectar to another 'house' bee that will process the nectar and store it.

If a worker finds a good source of nectar, she will tell her sisters in the hive where it is, using a formalised communication language in the form of a dance, and then give some of the nectar to other workers who attend the dance.

The other workers are thus informed of the position of the source of nectar and its flavour and sugar content. They are then in a position to visit the source of nectar and bring more back to the hive. In this way a colony can recruit many bees to fetch nectar and pollen from a local source that provides rich pickings.

How much honey?

In most years a colony can produce around 18kg (40lb) of harvestable honey.

Yields will vary from year to year and are dependent on weather – particularly the amount of rain, daily temperatures – not too hot or too cold during the summer months and any disease

It is said that a honey bee would have to fly the equivalent of six times around the Equator to collect enough nectar to produce 1kg of honey.

Worker bees exchanging nectar. The action of sharing and passing food between bees is known as trophallaxis.

Bee dance

The interior of a hive is pitch black except for a small area of light around the 'front door'.

Everything that goes on in the hive is done by touch, feel or smell. Bees clearly don't see each other inside the hive, so any house cleaning, feeding and other tasks have to be achieved in the dark and by touch.

How then do they communicate what they see outside, when foraging, to the bees inside the hive in the dark?

To advertise the source of nectar or pollen she has found, the returning worker bee performs the now famous 'waggle dance' which conveys information on the location of a food source in relation to the hive, the position of the sun and flight distance.

The returning worker bee will 'waggle' her abdomen whilst moving in a straight line, then she will turn and move in a semi-circle back to her starting point.

She will repeat this action again only next time she will return the other way. The straight line indicates the direction of the food source in relation to the sun's position.

As a result of the food sharing (trophallaxis) that occurs in a honey bee colony, all the bees in one colony will tend to have a similar odour.

Bee dances

Perhaps one of the most spectacular ways bees communicate is by the bee dance. The bees create a 'dance floor' on a patch of comb within the hive. This comb is left with minimal larvae or stores so its surface can vibrate if a bee shakes it.

Waggle dance

When a bee returns from a forage site that is a good source of food, she will go to the dance floor and perform a 'waggle' dance. In this dance she will shake her abdomen and align her body in a particular direction. She will then do a circle to the right and return to the same place and repeat the dance. She will then circle to the left and do the dance again until she has fully communicated her message. This dance will tell other bees who come to observe the dance, how far and in what direction the food can be found. It was not until the 1940s that the dance was understood when it was discovered that the direction to the vertical on the frame equates to the direction of flight relative to the

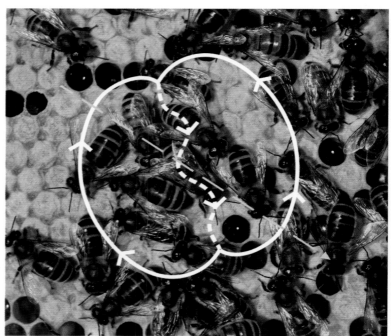

sun and the intensity of the waggle relates to the distance to fly. The more vigorous the 'waggle' is the closer the source of food. The single dance gives both direction and distance from the hive.

Remember that all this is done in the dark inside the hive so the potential recruits have to touch the dancing bee to determine the direction and distance to travel. Research is still going on to understand the dance; while the interpretation of direction is fairly clear the measurement of distance by the bees is not fully understood. It may relate to the time of flight or the number of markers (such as trees) passed or the amount of energy required to get there. Whatever it is, it works for the bees and once a bee discovers a rich source of food it will only take a few minutes before others from the same hive arrive to take advantage of the food.

The moral to the tale is do not leave any honey or comb exposed when keeping bees, for once one bee finds it, soon after there will be many hundreds more trying to get to the honey!

There is some dispute about the exact mechanism that scout bees use to ensure that other bees find the source of food. Some say that the bees rely only on the dances to find the source whereas other believe that the original scout bee leaves the hive and then guide a group of followers to the source. The role of the waggle dance is to ensure that the followers are prepared (have enough food) for the flight and leave the hive in approximately the right direction to be able to follow a scent trail left by the scout bee.

This is just one of many dances used by the bees to communicate information.

Round dance

If the food source is near to the hive then giving directional information is not so necessary and another dance is used, called the 'Round dance'. In this dance the forager will circle in one direction (say clockwise) and then reverse and circle in the opposite direction. This dance will go on for some time and is

Top: Worker bees in the dark feeling their way over a frame of sealed brood. The round white dots in the image are empty cells. This picture has been manipulated using software to look like an infrared image.

Bottom: Worker bees gorging on some honey spilled whilst shaking off a frame of bees containing unsealed honey.

This can occur quite easily especially with fresh honey of a high water content when closer inspections of the frame are required.

The round dance

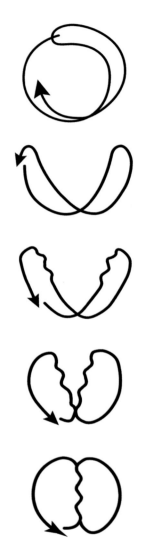

The waggle dance

Above: The transition from the round dance at the top to the waggle dance at the bottom.

Right: The dance floor where a returning forager will perform one of the dances described.

considered a very effective way to recruit followers. The outcome is usually many bees leaving the hive and circling about looking for the food source. The cross over from the round dance to the more complex dance to indicate a food source more distant seems to vary with the breed of honey bees. It is generally between 30 and 100 metres from the hive. Interestingly there is a third dance called the 'Sickle dance' which is used as a transition between the round and waggle dances.

DVAV dance

The dorsal ventral abdominal dance is performed by a worker whilst grasping onto another worker. It appears that this excites other foraging workers and encourages them to come to the dance floor to observe a foraging dance. The dance is also used on the queen. In this case it appears to prevent the queen from trying to destroy developing queen cells prior to swarming. It appears to excite the queen so that when the vibrations stop she is ready and willing to leave the hive with a swarm of bees.

Jostling and spasmodic dance

The jostling and spasmodic dances are performed by foragers and both seem to be used to encourage other foragers to come to the dance floor and 'observe' a dance about food resources. In the jostling dance a forager will return and knock into other bees causing a generalised disturbance. The spasmodic dance is also done by a returning forager. It involves waggling her abdomen but is short lived and is interspersed with unloading her crop of food.

Buzzing run dance

The buzzing run dance is also done by scout bees but in this case it is done just before a swarm leaves the hive. It disrupts the other bees in the hive and can be seen as a signal to drive the bees out to form a swarm.

Shaking dance

The shaking dance is done by bees that signal to other bees that they would like to be groomed. During the dance the bee bows her head and shakes her abdomen, Other bees are attracted and start to groom her, probably because there is some parasite causing some irritation.

The worker

From the moment a worker emerges from her cell, her work in the colony begins.

For the first three weeks of her life she stays mainly within the hive, cleaning cells, nursing the brood and feeding them.

Her wax glands will start to secrete wax which she moulds into new cells or repairs damaged parts of the comb. Towards the end of her stay she will take on foraging and guard duties.

Left: A well laid-up frame of mostly capped over worker brood. Notice a few domed drone cells around the outer edges of the comb.

The hive

The hive provides the space in which a colony of bees live. In nature this would be a hole in a tree or any cavity where the bees are able to have a single entrance that they can easily defend. A lot of research has been done to determine the nature of the cavity that a colony will look for when trying to find a new home. Although the size of the cavity chosen varies, usually between 40 and 80 litres, some much smaller and others much larger have been found.

A 'traditional' hive

The hive pictured is a WBC invented by and named after William Broughton Carr.

It is a double-walled hive with an external housing that splays out towards the bottom of each outer lift covering a standard box shaped hive inside.

The WBC is the 'traditional' hive as represented in pictures and paintings in the UK.

The WBC has lost favour with many beekeepers as it is a little more difficult to manipulate, having double walls, however this extra insulation may be of benefit in colder climates.

Note how easily adjustable the entrance is with its two slides that may be opened or closed to suit the time of year.

The history of the hive

Before the 18th century, beekeepers used hollowed out logs, skeps (tea-cosy shaped hives made from straw) or clay pots to house colonies of bees. These designs were fine for the bees to develop a nest but when the beekeeper wanted to remove honey the whole structure of honeycomb built by the bees had to be destroyed and often resulted in the colony being killed.

During the 18th and 19th centuries much research was carried out on the design of beehives that enabled honey bees to be kept in specially constructed boxes in such a way that the internal workings could be observed and the bees could be inspected without the need to kill them. A crucial finding was the discovery that in a honey bee colony the bees would leave corridors between parts of the nest to allow them to visit all parts of the hive.

In 1851 an American preacher called **Lorenzo Langstroth** patented a hive design that comprised a rectangular box with wooden frames inside that allowed the bees to build honeycomb on the frames and also ensured that the bees would not build honeycomb that connected the frames. This meant that the frames could be removed without the honeycomb being destroyed and then returned to use. This patented design, known as the Langstroth hive, still exists and is the most common hive design across the world. Since then many other hive designs have been developed to further improve on the basics of the Langstroth hive and, throughout Europe there are many different designs in use today.

A modern skep made to the traditional design and still in use. The skep and its base are placed on a hive stand to allow the beekeeper to more easily observe the colony.

In the UK the most popular designs are:
- Modified National
- Modified Commercial
- Deep National
- Smith
- WBC
- Langstroth
- Dartington Long Hive
- Beehaus

Recently there has been a large increase in interest in so-called sustainable hives. These include the:
- Top Bar hive
- Warré Hive
- Sun Hive

A nucleus hive – referred to by beekeepers as a Nuc. This is for moving bees or as a starter colony of five frames containing a queen, three frames of brood and two frames of food.

From the beekeeper's point of view **the hive design is considered to be of great importance** and beekeepers may argue for hours over the pros and cons of a particular design. However, the honey bee has simple needs and will happily establish a colony in any of these designs so long as:

1   The internal volume is large enough
2   The entrance is small and can be defended easily
3   There are no other access points other than the entrance
4   It is weatherproof providing some protection from the cold and damp.

When you are starting out as a beekeeper and acquiring your first hive, ensure that the needs of your bees will be satisfied, listen to the discussions of fellow beekeepers and then make up your own mind.

Key components of the hive

The majority of hives are designed in a similar way and comprise the following components:

Stand

The stand is **under the hive** and holds it off the ground. The stand must be strong and well built. It can consist of just two building blocks, two timber beams or a purpose built framework. It is very important that the stand supports the rest of the hive so that it is level. A hive that is fully laden with bees and honey can weigh up to 100kg so it is important that the hive stand is placed on hard ground that will not subside. This can be achieved by placing a concrete paving slab under the hive stand to distribute the weight.

Floor

The floor is **the base of the hive**. In the past this was a solid wooden base with wooden bars on three sides leaving one side to act as an entrance to the hive. There are many designs available and most now contain a wire or plastic mesh underneath that ensures debris drops out of the hive but is fine enough to prevent bees passing through. This mesh also allows Varroa mites (predatory mites that can devastate honey bee colonies) to fall out of the hive when they fall off a bee.

Top: A solidly built western red cedar stand together with the alighting board.

Centre: The stand is now fitted with a 'Varroa' floor – a wire mesh that allows debris to fall through to either a tray or the ground underneath so that it does not contaminate the hive.

Bottom: A small entrance block has been added to the floor unit.

Most floors provide a **small entrance to the hive** for the bees to leave and return. The entrance can be as small as 6mm deep and 70mm wide but most designs allow the entrance to be widened during the summer months when there is a higher rate of activity in the hive. Small entrances are good as they help the bees defend the nest from attack from other bees, wasps and other creatures such as mice.

It is very important that the hive has adequate ventilation because, while the bees keep the colony warm and when they process nectar into honey, water vapour is generated which can cause condensation and the build up of mould in the hive. The open mesh floors provide sufficient ventilation without the need for a large entrance.

A brood box on the stand and floor with frames placed in the 'cold' way (page 36).

Brood chamber

This is **the heart of the colony** and is the area where the queen lives and new brood develops. Most beekeepers place a 'queen excluder' above the brood chamber. This is a device that prevents the queen (and drones) from passing through into other parts of the hive and raising brood there. The brood chamber is the most important part of the hive as this is where the colony raises new bees and where foraging bees bring nectar and pollen back to feed larvae and for storage.

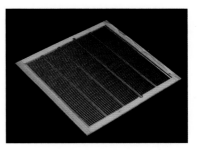

A steel queen excluder. This design used to be very popular, but framed wire extruders are now common as the do not get stuck to the top bars.

Queen excluder

This is a flat sheet, either made from plastic, zinc or framed and made of steel. It has holes in it that allow workers to pass through but **prevent the queen gaining access** to other parts of the hive. The objective is to keep the queen in the brood chamber and encourage the workers to store surplus honey on the other side of the queen excluder. In this way the honey crop removed by the beekeeper remains clean and free from debris, such as pupal cases left over from breeding new bees. The best design is made of rigid steel rods mounted on a wooden frame.

Supers

Supers are boxes that tend to be **less deep** than the brood chamber and correspondingly hold smaller frames. They are smaller because the bees fill them with honey and, when full, they can

A super has been added on top of the brood box above the queen excluder.

be extremely heavy. A National super will hold up to 15kg of honey whereas a National brood box would hold up to 25kg of honey and is difficult to lift.

Supers need to be removed when the colony is being inspected and if they are very heavy can be dangerous to lift and may cause back (and other) problems for the beekeeper.

Crown board

The crown board is a **framed sheet of plywood** or similar material that goes over the top of the colony. It defines the limit of the colony. There are many designs of crown board: some have holes in the top to aid ventilation while others are modified to serve other purposes, such as acting as dividers when the beekeeper is controlling the process of swarming.

Roof

The roof fits over the crown board and **provides weather protection to the hive**. It is quite heavy and ideally built with deep sides so that it is unlikely to be blown off. The top is usually made of a metal sheet that stops water penetrating. There can be insulation inside the roof to reduce heat loss; this is particularly helpful in the winter when heat retention is important.

A crown board has been added above a super and both brood and super boxes turned to the 'warm' way.

A deep sided roof with galvanised cover

Frames

The design of modern hives relies on bees making the wax comb on frames that can be removed without killing the bees or damaging the frame. The most common hives use a frame with a top bar, two side bars and a bottom bar that provide a convenient rectangular area for the bees to produce comb. However, some types of hive, such as the Top Bar hive and the Warré hive provide only a top bar from which the bees will hang the wax comb. In all cases the spacing between the frames is designed to encourage the bees to make a single sheet of comb on each frame and, hopefully, not to join adjacent sheets.

It is a good idea to purchase the **wooden frames** from specialist beekeeping suppliers, as the dimensions need to be accurate and the shape of some of the pieces is complex. The frames are supplied for specific hive types so that they fit with the right toler-ance. The components of the frames are then nailed together to provide a rigid framework to hold the wax foundation. Some frame designs are self spacing so that when they are placed inside the hive they have sufficient gaps between them to allow the wax cells to be built to the right depth and still allow a space through which the bees can walk. Other frames are designed to hold plastic spacers that achieve the same result.

Plastic frames are available for most hive designs. These are either designed to unhinge to receive a sheet of wax foundation or include plastic 'foundation' built into the frame. This latter

Parts of a hive

There are many parts to a modern beehive. For a National shown here each box is made up of four sides, screws, many nails and a pair of frame runners. There are eleven frames in the brood box and ten in the supers, each consisting of six pieces that are nailed together holding a sheet of wax foundation.

The special Varroa floor, roof and hive stand are perhaps a little more complicated to build.

The WBC has an inner set of boxes as well as the outer sloping sided boxes and is generally more difficult to construct.

If you have the time, a few hand tools and modest skills it can be very satisfying to build your own hives either from a flat pack hive kit as seen below or from hive plans available on the internet.

A complete hive showing (1) the hive stand; (2) floor with an entrance block; (3) a brood box with frames; (4) a wire queen excluder; (5) two supers, both with (6) frames; (7) a crown board and (8) a flat roof with a metal cover to protect the wood and keep the bees dry.

Hot or cold?

Like humans, bees need some fresh air in their homes.

To allow air to circulate freely within a hive you can set the frames in what beekeepers call the 'cold way' – at right angles to the hive entrance.

Use the cold way if your hives are in a sheltered spot and you have a partner helping you whilst working on a hive.

To reduce air circulation, use the 'warm way' with frames parallel to the hive entrance – this is best for an exposed or cold site and provides a better way for you to manipulate a hive, from the rear, if you are working alone.

type needs to be treated with a wax wash or covered in sugar syrup to encourage the bees to accept the foundation and build wax comb on the frame.

Foundation

The foundation consists of **a thin sheet of beeswax** that is embossed with hexagons on both sides to provide the bees with a substrate on which to build wax comb. The foundation is fixed into the frames and when placed in the hive encourages the bees

to build comb on the frames rather than in other places. Using this method results in a series of wooden frames that can easily be removed to inspect the colony and then replaced in their original positions. Wax foundation is made from beeswax reclaimed by beekeepers and supplied to companies that clean and purify it before forming it into foundation. It can be supplied 'unwired' or 'wired'. Because wax foundation is thin and not very strong it may sag in the frame; thin wire is embedded into the foundation to provide extra strength. Once the bees have drawn out the foundation and created hexagonal cells on each side the sheet becomes far more rigid and is able to maintain its shape.

Above: Wax foundation is being drawn out by workers to form the typical hexagon shape of the cells.

Centre: frame 1 is a National super frame fitted with wired wax foundation. Frame 2 is a wired standard National brood frame. Frame 3 is a wired 'Deep' National brood frame.

Wired foundation can also be useful in the frames that are used to collect honey. The most common way to remove honey from the wax comb is to place it in an extractor (a device that spins the frame in a similar way to a spin dryer); the force exerted on the wax in an extractor can be such that without the wire the wax comb may break and detach from the frame.

It is not always necessary to use foundation as a substrate for the bees to make wax comb. Some beekeepers will allow the bees to make their own substrate by providing a bead of wax along the underside of the top bar. The bees will use this bead as an anchor and draw wax comb down along this line.

This method is popular with beekeepers that use top bar hives and also beekeepers that make honey crops from oilseed rape. This kind of honey crystallises quickly in the comb and can be difficult to remove. Once crystallised the honey is removed with the wax and then separated at a later stage.

Beespace

Beespace is the **gap between any two components** inside a bee hive that is not either filled with wax comb or sealed with propolis to provide no access. All modern hives recognise the importance of beespace in their design. The normal figure given for beespace is between 6mm and 8mm and the frames in every hive are designed so that this gap exists all around the frame.

The sides of the frame are **one beespace** from the side of the hive and there is one beespace between the top of any box and the next box above it. Beespace is an essential component in the design of any removable frame hive. If this is not maintained between every part of the hive the bees will stick the components with propolis and build wild comb in gaps that make it difficult to inspect the colony without damaging the comb and annoying the bees.

When considering various hive designs it will become apparent that some are described as 'top beespace' and others as 'bottom beespace'. This is because the internal components of the hive must retain beespace whilst the outer box must be 'bee tight' with no gaps.

Top beespace designs result in a 6–8mm gap between the frames in the box and the top edge of the box, whereas bottom beespace designs have the frames and the top of the box flush, with beespace provided at the bottom of the box. The many debates about which is best that suggests that there is little difference

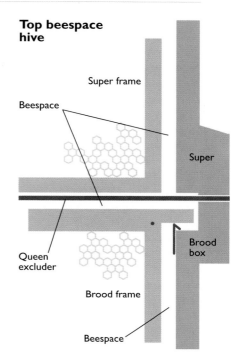

Top beespace hive

Super frame

Beespace

Super

Queen excluder

Brood box

Brood frame

Beespace

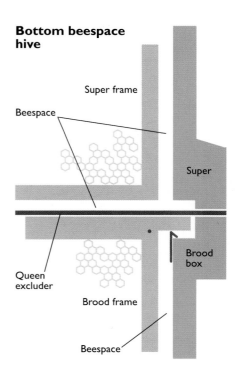

Bottom beespace hive

Super frame

Beespace

Super

Queen excluder

Brood box

Brood frame

Beespace

Hive materials

Hives have traditionally been constructed of wood to replicate the bees' natural home in a tree. The choice of wood has been western red cedar, pine, or sometimes plywood.

Red cedar is probably the best material for constructing a bee hive as it weathers well and will naturally resist mould.

Cedar takes on a lovely soft grey colour after a few years exposure to the elements and honey bees certainly seem to like it!

In recent years hives made from injection moulded high density polystyrene have become available.

between the two. For a new beekeeper it is better to use top or bottom beespace as dictated by the hive design chosen.

Beespace is approximately the height of an adult bee. By preserving beespace throughout the hive bees are able to visit all areas and ensure the colony is kept clean and that any predators are unable to find a safe haven in odd corners. It is interesting to note that there is one place were the gap between two components is twice beespace and this is between adjacent frames holding brood. This is because the workers need to tend the brood and it is necessary for workers to work 'back to back' in this area.

Hive materials

Most beehives are made of wood, primarily because it is easily machined to the dimensions required, and provides protection from the environment. As a natural material, wood is quite acceptable to colonies and when cedar is used, the material is naturally rot-proof and will last for many years. Beehives can be painted but only on the outside; the inside is better coated with

A Scandinavian hive made from high density polystyrene. High density polystyrene is a proven, durable and recyclable material that is now becoming acceptable for bee hives.

Polystyrene hives have been used in Scandinavia for many years where temperatures can range from -35°C in winter to +35°C in summer. This lightweight material keeps the bees warm and dry in winter and cool in summer.

The hive shown has been securely strapped together to prevent any damage during winter storms.

propolis and wax by the bees. If microporous paint is used, the paint will remain on the hive without bubbling and can be refreshed quite simply following a quick clean. The application of insecticides to the hive to protect the wood from boring insects (woodworm) should be done with caution, as many wood preservatives are toxic to bees. Before using any preservative you should always ensure that it is not harmful to bees.

Besides wood, beehives can be made of many other materials as long as they are weatherproof and not harmful to the bees. Plastic and expanded polystyrene hives are available and these have the added advantages that they are light and provide better insulation from cold weather. There are some drawbacks in that they can be more difficult to sterilise if there is an outbreak of disease in the hive while the light construction can mean that the material may be damaged when disassembling the hive as bees often stick components of the hive together.

When purchasing hives, a beekeeper must weigh up the pros and cons for his or her particular situation. Cedar wood components are expensive but will last for many years whereas expanded polystyrene hives are less expensive but may not last as long. When starting beekeeping it is always best to join a local group and ask for local advice. Generally it is better to use hives of the same design as your colleagues so that you have their expertise to draw on in the event of a problem.

Types of hives

Modified National Hive

The Modified National hive is probably the most common hive used in the UK. It is square with external dimensions of 460mm and the brood chamber is 225mm deep. The internal volume is about 35 litres, which may appear small compared with natural cavities. The National, like many hives, is designed so that the bees will store honey in the supers, each of which provides an additional volume of 22 litres. Therefore with one super on a National Hive brood box the hive provides a capacity similar to that which a colony might try to find in the wild. This method of

Hive roof styles

Hive roofs come in a variety of styles, from the shallow flat roof to the deep flat roof useful in more exposed areas and pitched roofs that look more in keeping with a traditional hive style.

Flat roofs are much easier for beekeepers as they provide a ready made stand when placed on the ground – always 'upside down' – to receive the supers when carrying out inspections.

An example of a modified National hive with a 14 x 12 inch brood box, complete with super and pitched roof sitting on a purpose made hive stand.

Brood frames

B.S. National frames are smaller than standard Langstroth and Commercial frames and have longer lugs.

Many beekeepers take the view that the brood box of the National is too small for the laying activity of modern strains of queen, so they operate a National with a brood box and one super. This is referred to as 'a brood and a half' system.

While this provides enough room for the brood, it also increases the number of frames that have to be checked over at regular inspections.

As a result, the National hive brood boxes are available in a 14 x 12 inch size which gives a brood size similar to that of a Commercial or Langstroth hive.

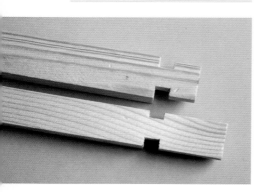

Top bars from a Commercial and a National frame clearly showing the shorter lugs on the top bar that the Commercial hive style uses.

comparison is rather complicated so the normal method used is to compare the number of cells available on a set of frames in the brood chamber. The frame sizes used in the catalogues are normally given in imperial measurements and relate to the outer dimensions of the frame.

In recent years the Modified National has been considered only just large enough for the colony sizes expected in the southern UK (say south of Manchester) but further north, where colonies do not become so large, it is considered quite adequate and remains a very popular hive.

The main characteristics are:

- The brood chamber (and the supers) have large rails along the side that help when lifting the boxes.
- The frames have long lugs that make lifting and manipulating the frames easier.
- It normally has bottom beespace.
- The brood chamber holds 11 frames with 50,000 cells.
- The frame size is 14in (355mm) wide and 8in (203mm) deep.
- The construction of the boxes is complicated and requires more components than other designs and is more expensive.
- It is single-walled with only one piece of wood between the brood inside the box and the outside environment. In cold climates this can mean the bees need to work harder to keep the brood warm.

Modified Commercial

The Commercial hive was originally designed for commercial beekeepers who wanted a brood capacity that was larger than that of the National along with a simpler design. The external dimensions are virtually the same as the National hive and many beekeepers who have Commercial brood chambers also use National Supers.

The Commercial hive does not have large carrying bars and therefore the internal dimensions are greater. The depth of the brood chamber is also about 25mm deeper than the National.

The frames are consequently larger and a full brood chamber with 11 frames will contain about 70,000 cells.

The simple construction and larger capacity have made this hive more popular in recent years and it is now frequently used by amateur as well as commercial beekeepers.

The main characteristics are:
- The brood chamber has small hand holds cut into the sides of the box.
- The frames have short lugs but consequently have a larger area for wax comb.
- It normally has bottom beespace.
- The brood chamber holds 11 frames with 70,000 cells.
- The frame size is 16in (406mm) wide and 10in (254mm) deep.
- The construction is simpler than the National hive and uses fewer components.
- It is single-walled.
- It can use either Commercial or National Supers.

Deep National

The Deep National is a variant on the National hive. The brood chamber has the same horizontal dimensions as the National but is 100mm deeper. This provides a large brood area and the frames hold about 72,000 cells. In effect it is slightly larger than the Commercial hive. It has gained popularity in the southern UK in recent years and it uses frames that are the same size as the Dartington hive (see page 47).

The chamber construction is more complex, like that of the National, and the frames are a bit more difficult to handle, but many beekeepers prefer the almost square shape of the frames and like its large capacity.

The main characteristics are:
- The brood chamber is large but has strong side bars for lifting and carrying.
- The frames have long lugs for easy removal.
- It normally has bottom beespace.

Hive choices

When starting out in your bee-keeping career, choose your hives carefully. Look at the hives your fellow beekeepers are using and seek their advice before making that first purchase.

When you have decided on the type of hive — stick to it for simplicity and ease of manipulation.

Bees will thrive pretty well in any style of hive, but from your point of view, mixing and matching hives of different styles or from other manufacturers can present problems. All National boxes will be similar but may have minor differences or vary slightly in size.

For most beginners, starter equipment should be bought as a 'made up package' as this provides a complete hive from one source that will fit together perfectly.

The same is true of frames that can be bought made-up with either wired or unwired wax foundation in many different sizes and styles.

Making up your own frames is fairly easy if you can handle a small hammer and tiny nails without banging fingers or thumbs!

Brood boxes

Special wood fittings may be purchased or made if you have modest carpentry skills to turn a Standard National brood box into a Deep National.

This additional kit allows you to start off with a standard set-up and convert at a later stage as your beekeeping skills and confidence improve to run a 14x12 inch or Deep National system with consequently many more bees.

Right: the Author inspecting a 14x12 or Deep National brood frame.

Above: Extra kit required to turn a standard National brood box into a deep or 14x12 brood box.

- The brood chamber holds 11 frames with around 72,000 cells.
- The frame size is 14in (355mm) wide and 12in (305mm) deep.
- The construction is complex as in the National hive.
- It is single-walled.
- The large frames mean that the wax comb may sag when out of the hive if they are not handled correctly.
- It can use National supers.

Smith

The Smith hive is popular in northern parts of the UK, particularly in Scotland. It is of simple construction but, unlike all the hives mentioned previously, the hive is rectangular and not square. The frames take the same size wax comb as the National but the lugs on the frames are shorter. Therefore the internal dimension of the brood chamber are the same as those of the

National but the overall dimensions of the box are smaller and the design makes for easier construction.

When Mr. W. Smith, a Scottish beekeeper, designed this hive, his intention was to create a small-scale Langstroth hive that could be easily carried to the heather moors. Those who use the Smith hive like its simplicity and find the volume in the brood chamber is adequate for the size of colonies they have.

The main characteristics are:
- Small outer dimension and not square.
- The frames use short lugs.
- It normally has top beespace.
- The brood chamber holds 11 frames with 50,000 cells.
- The frame size is 14in (355mm) wide and 8in (203mm) deep.
- The brood chamber is simple to construct.
- It is single-walled.
- It can only use Smith supers whereas the National and Commercial hives can use National supers.

WBC

The WBC hive is named after the Reverend William Broughton Carr who designed this hive in the late 19th century. Its main feature is an internal construction of boxes where the bees reside and an outer set of 'lifts' that provide weatherproofing and an attractive shape. It is not as common now as it used to be but is the archetypal hive that can be seen in pictures of many country gardens.

The double-walled design creates useful insulation and weatherproofing and means that the internal components can be made of lighter materials. However it is more expensive to purchase and more complex to disassemble when inspecting the colony, while the design of the internal parts encourages the bees to stick the components together. Those who have this design of hive enjoy them and place them in prominent positions in their gardens. Recently the overall design has been used to produce good-looking compost bins, so not all beehives are just for bees!

Single or double?

Most beekeepers end up using single-walled hives for simplicity.

Although double-walled hives look attractive in a garden setting and are probably more suitable in colder areas of the country they are more complicated to manufacture and more expensive to purchase than the Langstroth, Dadant or Modified National hives.

Double-walled hives are essentially two hives in one with inner hive boxes and outer lifts and therefore take more time to inspect.

A WBC hive marked with an 'Ace of Clubs' symbol that may help returning bees find their correct hive in a large apiary.

Frame spacing

No matter what hive design you choose, the importance of correct frame spacing within the brood and super boxes should be carefully considered.

Frame spacers are small plastic parts that slip over the lugs of a brood or super frame to ensure correct bee space is maintained between each frame.

Hoffman 'self spacing' frames are common and many beekeepers use them in their brood boxes. These have a wider and shaped side bar to the frame that allows for correct bee space when frames are butted together in the brood box.

Castellated metal runners with predetermined slots ensure correct spacing of frames in supers, where wider spacing encourages bees to build deeper cells and produce more honey.

Manley style frames, with very wide side bars, are used in honey supers to create a wider spacing without using castellated runners.

The main characteristics are:

- Double-walled giving better winter protection.
- Expensive to buy but attractive.
- Complex to manage when inspecting the bees.
- It normally has bottom beespace.
- The brood chamber holds 10 frames with 45,000 cells.
- The frame size is 14in (355mm) wide and 8in (203mm) deep.
- The outer dimensions of the brood chamber and supers are not square.
- It must be used with WBC supers as others will not fit.

Langstroth

The Langstroth hive is probably the most popular design in the world and is used in many countries. It is the standard hive in North and South America and Australasia. This is the original design patented by the Reverend Langstroth in 1851 and is reputed to have been based on the dimensions of a case of champagne. It is rectangular, rather than square, as are the frames that fit inside to hold the wax.

The brood chamber is larger than the National hive but smaller than the Commercial. It tends to be a little cheaper than the other 'British' hives because it is produced in such large numbers across the globe.

Modifications and developments in the US are readily available in this country as well. There is a variant known as the Jumbo Langstroth that has deeper frames and gives a much larger brood chamber.

The main characteristics are:

- It is single-walled and easy to construct.
- It is rectangular not square.
- It normally has top beespace.
- The frames have short lugs increasing the internal dimensions.
- The brood chamber holds 10 frames with 62,000 cells.
- The frame size is about 18in (457mm) wide and 9in (228mm) deep.
- The Jumbo Langstroth frame size is about 18in (457mm) wide and 11in (289mm) deep with about 75,000 cells.
- It must be used with Langstroth supers.

Dartington Long Deep Hive (DLD hive)

The Dartington Long Deep hive was developed by Robin Dartington in the 1970s to reduce problems with the National hive for hobby beekeeping – in particular unsafe weights, need for extra equipment for controlling swarming.

Robin changed to 14x12 inch frames (first introduced in the 1900s) to avoid using 'brood and a half' – and doubled the length of the deep brood box to both the 'swarm' and the 'parent' in the same hive after artificial swarming. The long box can be used for holding one colony of any size between two insulating dummy frames or to separate different queens just by dropping in a division board and opening the rear entrance.

To reduce lifting, the supers are half-length, called 'honeyboxes', each weighing less than 8kgs. The honeyboxes can be spread

Bees on the roof

Bees are very tolerant of man's intervention and where space is limited beekeepers are looking for opportunities to keep bees.

As you become more experienced you may wish to keep bees at home and with your neighbours' agreement, housing them on a flat roof that can be easily accessed has its merits.

This style of urban beekeeping is popular in many large towns and can produce a larger honey crop from a wider variety of flower sources than the more traditional rural apiary.

Rooftop beekeeping is not for everyone and needs careful consideration, for your neighbours' and your own safety.

A classic design

An early style of Dartington hive, shown here painted white, alongside two Modified National hives for comparison.

The Dartington Long Hive's body provides the same volume as two Standard National bodies plus two honey supers placed side by side.

At first sight it may appear a large hive – but the overall volume when its four honey-boxes are in place is in fact less than a National in mid-summer, when two brood boxes and four supers are often needed to store all the honey.

Above: a 'Beehaus' hive, built along the lines of a Dartington Long Hive, but fabricated in plastic with metal legs.

Centre: an early Dartington Long Hive, alongside two National hives for comparison.

down the length of the hive or stacked over the brood at one end, with the split roof then being at different levels, equivalent to a Deep National with a back extension.

A dividable longer box with an extra optional rear entrance was called a 'combination hive' in the late 1800s as it allowed honey to be stored both beside and above the brood, and was able to house both a colony and a nucleus. It fell out of fashion as too cumbersome to move for pollination but that objection applies less today as few hobbyists move occupied hives.

The design details of the modern Dartington re-introduce the roof eves and tunnel entrance of the early moveable frame hives. The overall design addresses 'safety, convenience and economy'.

Main characteristics:

- The 'all-in-one' design creates a stylish hive to enhance the garden – no piles of dirty loose boxes in view or taking up the shed.
- The number of 14x12 inch deep frames for brood including stores beside the brood nest is unrestricted, typically nine for winter, up to 15 at the colony peak, up to 21 when the colony is divided.
- A divided colony is easily united under a new queen by removing the division board - and in autumn the colony is reduced again to nine frames for winter.
- Two additional 'carry-boxes' store the 12 deep frames that

are added and subtracted each season – and form temporary nucleus hives for rearing extra queens.

- Additional honey boxes, each holding five shallow Manley frames, can store honey above the brood during honey flows.

- The 12 deep frames removed before winter can be extracted in a Lega 9-frame radial extractor using the supplied tangential screens. Alternatively, the removed frames can be removed in the carry boxes and any honey transferred back by the bees into honeyboxes holding shallow frames.

- The longer hive avoids need to lift honey boxes high-up – the attached legs lift the body to a comfortable height and avoid stooping when inspecting the brood.

- Construction using plywood and softwood requires only gluing and screwing of rectangular components without the need to make joints. Components can be cut from standard timber sizes available from Wickes stores. A booklet, '*Construction Information*', for making hives at home is available from bee book suppliers.

- Recommended use of garden paint improves appearance, avoids wood getting wet with risk of wind-chill and prevents deterioration. Regular maintenance only requires an extra coat of water-based paint applied on a fine day in the apiary.

Top Bar

Top Bar hives are a relatively modern innovation that was designed originally for use in Africa. In recent years this design has gained popularity in the UK and other parts of Europe because it is seen to be a more 'natural' way to keep bees. The hive is of long hive design but the sides are sloped outwards, similar to the shape of a boat. The top of the hive comprises parallel slats of wood that abut and normally have a slot in the underside filled with wax to encourage the bees to build wax comb from this site. This design is considered to be more natural because the bees are not forced or encouraged to build honeycomb in a frame that has a preformed foundation. In the Top Bar hive the bees will create their own wax comb, hopefully fixed to the wax bead on one of the top bars. The shape of the hive encourages the bees to leave the wax foundation free from the

Wild bee nests

In the wild, bees will naturally build wax comb in an approximately semi-circular profile.

The sides of the Top Bar hives in the lower picture are designed to roughly mimic this shape.

Top Bar hives in the Kenyan jungle. They are very popular in Africa.

sides of the hive and thus it is easier for the frames to be removed.

The main characteristics are:

- The hive is long and cheap to construct.
- There is no standard top bar hive.
- The capacity of the hive entirely depends on the size of the constructed box.
- It has no supers and any honey removed comes from the same chamber where the bees raise brood.
- The frames only comprise a top bar.
- Inspection requires very careful handling of the frames.
- Comb often becomes stuck to the sides of the hive and can be broken when inspecting the bees.

Warré

The Warré hive was designed by a French beekeeper called Abbé Warré in the early 20th century. The idea was to design a hive where the manipulation of the colony was kept to a minimum and the hive could increase in size, as the bees needed more space. The hive consists of a series of equal-sized chambers that are square and have a volume of about 20 litres.

The top of each box is a series of eight wooden bars similar to the ones in the Top Bar hive except that these do not abut but leave a gap for the bees to pass through. As the colony develops and becomes larger, another box is placed below the first to give the bees more room. This continues as the colony gets larger and the theory is that the colony normally develops its comb downward while the colony gets larger and stored honey is kept in the upper chambers. There is a crown board and roof on the first box to provide weather protection.

The main characteristics are:

- There are no frames only top bars.
- All boxes are the same size.
- In effect it has top beespace.
- The volume of the hive can be increased by lifting the whole colony in its existing hive and placing a new box underneath.
- Honey is extracted from the frames in the upper boxes that have been used to raise brood.

A Warré hive, complete with a glass observation window in one of the boxes.

- It is difficult to inspect the brood and check for disease.
- It requires strength to manage as the whole hive needs to be lifted when adding another compartment.

Natural beekeeping

Recently there has evolved a strong following for those advocating Natural Beekeeping as a better method of keeping colonies of honey bees. Natural Beekeeping is a broad term for a range of beekeeping practices that vary from providing a cavity where bees are left alone to develop and (hopefully) survive until ready to swarm and move elsewhere. At the other limit to Natural Beekeeping there are those that keep bees in Warré or top bar hives and minimise the number of inspections. The belief behind these methods of keeping bees is that intervention by the beekeeper is both stressful and harmful to the colony and should be minimised. Unfortunately, whilst this sounds like good practice, it can result in colonies swarming and disturbing neighbours and it can also mean that colonies can suffer and die because manipulations or medication has not been applied to control disease.

There is much debate between conventional beekeepers who use hives with moveable frames and those who use Top Bar hives about the best way to manage bees. This debate will continue indefinitely as in the end all beekeepers care for the animals they have and, while there are advantages in each way for the bee, there are also problems with each approach. Only bad, incompetent or ignorant beekeepers will treat their bees in a way that is harmful to them.

Because of the difficulties facing honey bees and the dramatic reduction in feral colonies since the arrival of Varroa in Europe and across the world, interest in keeping bees in a similar environment to the wild has increased. Honey bees have not only suffered because of the prevalence of Varroa but, like all pollinators, they have suffered because of a lack of habitat as farms have become larger and more mechanised. These days, pollinators often appear to exist in greater numbers in urban areas than they do in the countryside!

A helping hand

There is plenty of scope for people to help bees, by providing nests, and the 'natural' beekeepers should be applauded for helping in this way.

However, if honey bees are to survive they need more than just a nest site; they need care from the beekeeper.

The hive type is not relevant but the care and attention that beekeepers give their bees is important. 'Let alone' beekeeping where the bees are left to their own devices is not sufficient to maintain a healthy colony. Any hive design must allow the colony to be inspected and cared for.

A Sun Hive with roof off, thereby exposing the semi-circular frames.

The history of bees

Bees have been around for many millions of years and have developed alongside the evolution of flowering plants but their ancestry can be traced back to very primitive insects. Understanding the background of honey bees helps to explain some of the peculiarities of their development, such as why they have such a narrow waist, and why only females are able to sting.

Honey stores

Capped-over honey stores will last a long time providing the moisture content is lower than 18–20 per cent.

In this image we see capped honey cells above some cells that are not yet completely filled with honey or pollen.

Uncapped honey stores are not sealed over with wax until the moisture content has fallen below 20 per cent as the honey will start to ferment and become useless for the colony.

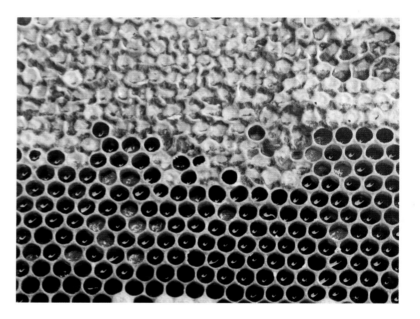

Capped honey above uncapped stores of honey and pollen.

The bees' ability to **collect and store honey** has been exploited by many animals that feel able to withstand the mass stinging attacks from the colony to steal some of this sweet liquid. Most animals that plunder bees' nests are well protected with thick skins often covered in fur. Humans have neither of these protections and have had to use thought and cunning to get past the bees' defences. It is only recently that humans have appreciated the important role of bees as pollinators. As a result methods of keeping bees have been developed that are to the advantage of the colony and their work as pollinators. Humans have also found ways to encourage bees to produce more honey than the colony needs so that the surplus can be taken by the beekeeper.

The ancestors of the honey bee

The honey bee as we know it has been around for about 40 million years without significant change. Its ancestry can be traced back much further, over 200 million years, to a more primitive insect, similar to the sawfly that exists today. The insect was a solitary insect and the female had a long ovipositor (egg-laying tube) that has a saw-like appearance. She used this to bore into the plants where she laid her eggs. These hatched and the resulting larvae continue to live on the plant. The larvae pupated inside the plant and then emerged as an adult, mated and the cycle was repeated.

This ancestor had a narrow waist that allowed the fly to bend its abdomen down and thus push the ovipositor into the plant. This modification was essential for the life history of the insect but had the unfortunate consequence that the fly was only able to ingest liquids. The stomach was in the abdomen and the narrow waist means that the tube joining the mouth to the stomach was very thin and would only admit liquids.

An ichneumon wasp.

Ichneumon wasps

Around 160 million years ago **a group of sawflies developed** the ability to parasitise other insects' larvae by using their ovipositors to lay eggs on or inside the unfortunate larva. The modern descendants are known as ichneumon wasps (often mistakenly called ichneumon flies). These wasps were also solitary and had a 'waist' between the thorax and the abdomen that makes their bodies far more flexible. This group of insects have developed further adaptations over the years and are now highly successful and able to attack insects much larger than themselves. The larvae develop inside the host but do not kill the host until they are ready to pupate into an adult ichneumon wasp.

Solitary wasps

Around 150 million years ago the ancestor of **the ichneumon wasp further developed into the solitary wasp**. This adaptation involved the ovipositor becoming a sting mechanism that is used by the adult to paralyse the host larva. Most solitary wasps will then drag or carry the unfortunate victim, alive but unable

A solitary Potter Wasp feeding on a Manuka (*Leptospermum scoparium*). These wasps lay an egg into a single small mud tube that they construct and after laying, leave a paralysed insect in the nest for the larvae and seal the tube over.

to move, to a rudimentary nest where the wasp will lay an egg on the larva. The egg hatches and feeds on the larva until it is ready to pupate.

Many solitary wasps have adapted to prey on one specific insect. They are able to sting all the nerve clusters in the victim and thus paralyse the unfortunate individual. In some cases paralysing the body is not sufficient and the wasp will then bite into the back of the head to paralyse the mouthparts of the victim. This specialisation and adaptation means that once the egg(s) of the wasp have been laid on the prey, they will have fresh meat from the prey whilst they develop. The prey eventually dies once the larvae pupate.

Solitary bees

Solitary wasps were the genesis of solitary bees about 120 million years ago. The **bees differ from wasps** in that the protein used by their larvae comes from plant pollen rather than the larvae of other insects. The development of solitary bees coincided with the evolution of flowering plants. The crucial relationship between bees, other pollinators and flowering plants was an essential requirement that allowed flowering plants to develop and evolve to become dominant over wind-pollinated plants. Although their role was important, bees were not the only pollinators, other insects such as butterflies, flies developed at about the same time.

Towards sociability

Over 60 million years ago bees began to show a degree of sociability. Colonies of solitary bees that gathered to share nest sites developed the ability to live as a single group in which each insect had a specialised duty. This was the start of the concept of a queen bee who lays all the eggs and workers who, although female, sacrifice the ability to lay eggs in order to care for the larvae of their queen and forage to collect food for the colony.

Top: A solitary digger wasp taking prey – a green fly, to its burrow to feed larvae.

Bottom: A solitary bee feeding on some Conference Pear blossom.

The major evolutionary development here is the differentiation of the gender of male (drones) and females (workers) that is achieved by males coming from unfertilised eggs and females from fertilised eggs. This has a major effect on the colony balance between male and female, moving from chance and approximately equal number of males and females to a greatly reduced number of drones. This adaptation means that the colony becomes dominated by productive females. The queen is the only one that needs to mate while the workers become the group that manages the colony. This imbalance in the male/female ratio and the removal from female workers of the need to procreate enables the social colony to survive and thrive. The semi-social bees that developed 60 million years ago were very similar in their organisation to the current colonies of bumblebees.

Then, around 50 million years ago, solitary wasps also started to become sociable. The wasps, which evolved at that time, are very similar to the ones we know today. However, although wasps can be a nuisance in August when they try to find sources of sweetness for food, at other times of the year they do a good job in the gardens collecting other insect larvae from plants to feed their own larvae. When the adult wasps feed the larvae the larvae produce a sweet substance as a reward to feed the adult wasps. The reason wasps annoy us in August is because their colony starts to break down and is not producing sufficient worker larvae to feed all the adult wasps. At the same time the colony produces future queen and drone wasps that will mate, and the fertilised queens then hibernate through winter to start new colonies in the spring. Meanwhile the old colony dies out and the nest is not used again. The wasps that cause problems in August are starving and will soon die as their old colony ceases to exist.

Wasps

Wasps often appear intoxicated in the late summer and autumn when they feed on apples and other fruit.

It is therefore sensible to warn children of the potential dangers of being stung and keep pets away from these 'stingy' but nevertheless useful insects.

Social wasps feeding avidly on a windfall apple.

The evolution of honey bees

Truly social bees evolved about 40 million years ago. These bees were able to survive as a colony for more than one year and could store honey to sustain the colony during times of a short-age of food. The social bees developed in Asia where their nests were out in the open.

There are still races of bees that live this way in tropical areas of Southern Asia. There are two main races: *Apis dorsata*, or the Giant Honey Bee, produces a single large honeycomb; and *Apis florea*, the Small Honey Bee, produces a series of parallel honey-combs for the colony. It is probably that these 'tropical bees' extended their range into Central Africa, for about this time the continents of Africa and India were joined together. Soon after the continents drifted apart and the bees in Africa were isolated from the other bees in Asia.

About 20 million years ago in both continents further development and adaptation occurred. In Asia a new species (*Apis cerana*) developed that build nests in cavities, giving further protection and the ability to move to more temperate climates.

Above: European honey bees (*Apis mellifera*) avidly feeding on late summer flowers.

Centre: migration routes of the honey bee from Asia to Europe via Africa.

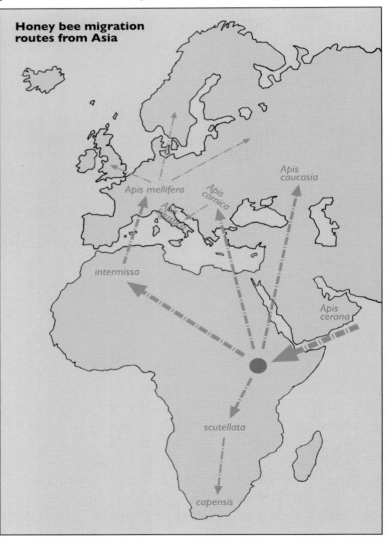

Honey bee migration routes from Asia

Apis caucasia

Apis mellifera

Apis carnica

Apis mellifera

intermissa

Apis cerana

scutellata

capensis

In Africa the species also developed into cavity nesting bees but unlike Asia no modern equivalent of the tropical bees with colonies in the open developed. Africa was dominated by one species (*Apis mellifera*) although there are many races or sub-species across the continent.

The evolution of the European honey bee

Our modern honey bee, *Apis mellifera* originally developed in Africa, and moved north into the Iberian peninsula and through the Middle East into Europe to produce what is generally known and recognised as the European honey bee. There are a number of notable subspecies within Europe that were originally associated with specific regions.

- *Apis mellifera mellifera* is known as the Black Bee. It is mainly found in western and northern Europe and western Russia and is naturally more able to survive in cold and wet areas.

The Black Bee

The northern European Black Bee (Apis mellifera mellifera) is the native bee of these islands.

This species has longer hairs on its abdomen and is said therefore to be able to better withstand our UK maritime climate.

The charity BIBBA helps bee-keepers raise native or near-native bees of this race that are generally more hardy than some others.

Apis mellifera races in Europe

Apis mellifera
Apis mellifera
Apis mellifera
Apis mellifera
Apis mellifera
Mellifera
Caucasia
Carnica
Ligustica
Apis mellifera

Apis mellifera mellifera is a modern example of the native European dark bee.

- *Apis mellifera carnica* is originally from eastern Europe and is adapted to the climate in this area. It can develop into a large colony in spring when the weather changes from very cold to hot but is able to survive as a small colony in the cold, dry winters.
- *Apis mellifera ligustica* originally comes from Italy and is adapted to the Mediterranean climate that does not get too cold in winter and in summer has an abundance of flora to feed large colonies.
- *Apis mellifera caucasia* comes from the mountainous region of the Caucasus. It is used to living in hard conditions and does not produce very large colonies. It is also slow to build up numbers in the spring as in its natural habitat the weather can sometimes take a long time to warm up while the flora can be slow to flower.

Until man started to move bees from one region to another these species remained relatively isolated by the mountains in the Alps and the Caucasus. In addition the European honey bee subspecies remained isolated from the Asian honey bees by Afghanistan and the Himalayas.

In the UK, honey bees, along with many species of animals and plants were unable to exist during the Ice Age but the bees returned as the ice receded and were well established before the land bridge between France and England disappeared. There is evidence from archaeological digs and peat bogs that the British Isles have been occupied by honey bees for at least 10,000 years and may have existed in the interglacial period before then.

Honey hunters

Early humans **did not keep bees** or manage colonies but would rob colonies of their honey whenever the opportunity arose. This was not a job for just anyone as the dangers from multiple stings were enough to ensure only the bravest (or hardiest) people would attempt the job. In the wild, honey bee colonies are found high in trees or hidden in well protected cavities. In these circumstances not only are there dangers from the bees themselves, but access to the entrance of the hive can also be difficult. Honey hunting is still practised today in some parts of the

A honey hunter taking honey from a nest while surrounded by angry bees is shown in this depiction of a rock painting from the Cueva de la Araña near Valencia in Spain, thought to be around 6,000 years old.

world, particularly with tropical bee colonies where the colony has built its honey comb in the open air. The honey hunters of Nepal are famous for their bravery as the bees from which they steal honey, the Giant Honey Bees (*Apis dorsata*) have powerful stings, and a honeycomb found fixed to high cliffs.

The origins of beekeeping

The idea of **hollowing out a log** to make a cavity suitable for bees to nest in evolved many thousands of years ago. The advantage was that the log could be hoisted into a tree to make it difficult for other predators to get to the colony but at the same time, the beekeeper could lower the log to the ground when it was time to remove some of the honey. This approach is still used in some parts of Africa.

The advantage of providing a receptacle for honey bees to build a nest is also that it is possible to move the colony nearer to your lodging to make it easier to collect honey when ready and to protect the colony from other beekeepers who might also like to take the honey!

Above: A log hive placed in a tree in equitorial Africa.

Left: A medieval illustration of a papal mitre depicted as a beehive, from Filips van Matrix's satirical work (1581).

There is documented evidence that the Pharaohs of ancient Egyptians and many **ancient civilisations hunted for honey** that was very highly valued. Beekeeping was well established in the UK before the arrival of the Romans and there were Celtic laws that referred to the ownership of bees. However, the

Romans gathered much information about beekeeping in their occupied lands and were able to spread good practice. So much so that many of the beekeeping methods practised by the Romans survived into the 20th century. The use of plant material to fashion hives was popular and the hives were daubed with mud or dung to make the surfaces smooth and waterproof.

The Roman laws on bees and beekeeping were very similar to those of the Celts. Bees were considered to be wild animals and anyone who captured a swarm became the natural owner. It was an offence to steal bees from the owner but if a swarm came from a hive and the owner lost sight of it, it reverted to being wild and could be collected and owned by anyone. In Europe bees were considered to be owned by the landowner who retained a bee ceorl (a freeman beekeeper working for the lord of the manor) to look after 'his' bees and provide honey and wax.

It is clear from writings that beekeeping, i.e. keeping bees in specially made containers and honey hunting (collecting honey from wild colonies) coexisted for many centuries in Europe right up to the Middle Ages and probably beyond. In Europe it was common to house colonies of bees in trees. These were often natural wild colonies but the beekeeper had cut a section out of the tree so that the wild comb could be accessed.

Monks became skilled beekeepers having realised that beeswax was an ideal material to make candles for monasteries and churches. Tallow, the alternative at the time, produced a smoky and smelly flame while beeswax burns with a bright, clear flame with virtually no smoke and became prized as a source of light. The added advantage of keeping bees was the production of honey and, when it fermented, a pleasant alcoholic drink known as mead that was also much enjoyed by the monks. As land ownership developed so too did beekeeping and individual farm workers began to keep bees as a source of sweetness and additional income.

Skeps and hackles

In Europe the beehive, the artificial cavity provided for bees to build their nests, was usually a skep. It was made from straw

Top: Bee boles at the Lost Gardens of Heligan in Cornwall. Boles can be cavities in walls or a free standing structure. They are built to house skeps in areas where weather conditions were less suitable for beekeeping.

Bottom: Straw 'hackles' placed over skeps, to help keep the bees dry in wet weather.

bound with blackberry bark or twigs woven into a conical shape and covered with mud to fill the gaps between the twigs. The base was open and the skep was usually placed on a wooden or stone base that included a small entrance for the bees.

The objective was to collect a swarm in spring and allow it to build up during summer into a large colony that could produce plenty of honey. The hives that produced most honey were placed over a sulphur fume pit to kill the bees and the wax comb with honey was removed for use and consumption. Sometimes the beekeeper would encourage the bees to leave their skep and move to a new one. This allowed the comb to be collected without killing all the bees and was known as 'driving the bees'.

These hives were not waterproof so a straw hackle was used to cover the skep and help to protect it from the worst weather. A hackle is simply a bunch of straw or hay tied near one end and spread into a conical shape and placed over the skep so that any rain falling on the hackle will just run over the surface and away from the skep.

Throughout Europe and the Middle East there were many variants on this basic device such as clay barrels, jars and hollowed out logs or purpose-built boxes.

Myths about bees

The Greeks, notably Aristotle, studied bees and came to a number of conclusions about their life. His work was considered to be so accurate that it remained largely unchallenged until about the 17th century. Then, during the 18th century, many European academics also studied the honey bee in detail and soon realised that some of the Aristotelian concepts were not valid.

It was discovered that:
- The largest bee in a colony was not a king but a queen (she laid eggs).
- Colonies of bees in the form of swarms came from other bee colonies and not from the dead carcasses of animals. (This idea was known as bugonia and is the source of the

Clay barrel hives used as beehives in Turkey and Kenya.

An illustration showing the concept of bugonia (from the Greek for 'ox birth'), whereby swarms of bees arose from the carcasses of dead animals. Rituals were followed that were thought to encourage the presence of bees.

'Tanging a swarm' – banging on a can to encourage the swarm to land.

A willow skep. This design remains almost unchanged from skeps used for centuries beforehand

image of a swarm of bees above the body of a dead lion that appears on tins of Tate and Lyle's golden syrup with the caption 'Out of strength came forth sweetness'.

◼ Bees do not 'collect' honey but process nectar to produce honey.

◼ Bees do not collect wax but produce it from glands under their abdomens.

◼ Bees do not live for six years. In summer most workers live for only six weeks.

◼ Bees do not carry stones in high winds to stabilise their flight. (The phenomenon that gave rise to this idea was probably the pollen that the bees carried on their back legs when bringing it back to the nest.).

Beekeeping distributes the European honey bee

The European honey bee was seen to be an excellent producer of a sweet substance (honey) by the Europeans. This was before the discovery of sugar cane and sugar beet, which now provide table sugar in large quantities. When European explorers began their great voyages of discovery from the 15th century onwards they took honey bees with them on their travels. Before this, the species *Apis mellifera* did not exist in North and South America, Australasia and Southern Asia. There were other bees and insect pollinators on these continents but these were either not as manageable or as effective in producing excess honey. The European honey bee has now been spread across the world with the help of humans, and as the world's primary producer of honey, is known to be essential to the pollination and subsequent development of countless crops worldwide.

In the 17th and 18th centuries the natural boundaries between subspecies were broken down as experiments took place to find the most productive and gentle honey bee for use by beekeepers. The characteristics of Italian and Carnican bees from central and southern Europe found favour with beekeepers and are the most common types of bee in many areas of the world.

In the early 20th century the 'Isle of Wight' disease wiped out many colonies in southern England. The disease was then thought to be related to Acarine (a parasitic mite on honey bees) but later surmised to be due to a viral attack. The bee population was supplemented by imports from across Europe and has resulted in the majority of honey bee colonies in England (and to a lesser extent in Scotland and Wales) no longer being pure-bred subspecies but a combination of all the European honey bee strains. Groups of beekeepers have tried to re-establish the black bee in England and increase its dominance but this is very difficult given the problems controlling the mating of honey bees.

The outcome has been that in the UK, especially in England, the majority of honey bee colonies are a mixture of all the European strains and the gene pool is quite diverse. This has a number of advantages: colonies are more able to adapt to the changing environment and climate and most colonies benefit from a degree of hybrid vigour. Many beekeeping experts would suggest that as local colonies adapt to local conditions they might ultimately be more suited to the local conditions than the original pure-bred subspecies.

Recent experiments across Europe have shown that locally adapted colonies of honey bees are both better honey collectors and more disease resistant. Despite this many beekeepers persevere in importing pure bred honey bees from other areas of Europe in the hope of establishing improved bees in their apiary. This brings dangers, as any importation of bees will increase the risk of introducing diseases or pests that were not present in the local colonies.

During the 20th century a monk at Buckfast Abbey in Devon (UK) was given the responsibility of maintaining the bee colonies at the Abbey. Brother Adam became an expert beekeeper and set about developing a 'perfect bee' for the Abbey. He travelled through Europe and Africa observing the races of bees and then established a breeding programme crossing various races in Devon. The outcome was the Buckfast bee that was particularly good at producing large quantities of honey and yet was calm and easy to handle.

Top: Buckfast Abbey in Devon.
Bottom: The dark bees that are typical of the Buckfast strain.

Brother Adam became world renowned for his work and after his death this work was taken up by other enthusiastic groups.

Straw skeps

Straw skeps were the precursor of the moveable frame hive.

To harvest the honey and wax, the bees were killed and the honey comb removed. To process the honey the comb was crushed and the liquid honey passed through a cloth, an inefficient way of collecting honey.

In the spring, swarms from surviving skeps were reinstalled in the skeps and the process repeated.

The Stewarton hive, designed by Robert Kerr in 1819.

Unfortunately this work did not continue at Buckfast Abbey. It is clear now that the Buckfast bee is not a new race of bee but the outcome of a complex breeding programme. However many beekeepers believe that the Buckfast traits are worth preserving and try to maintain these in their colonies.

The development of hives

During the 18th century a number of **hive designs were developed** to reduce the need to kill bees in order to access their honey and wax. One of the first was the development of a straw 'super' which was a smaller skep that sat on top of the skep that held the bees. The lower skep had a small hole in the top to allow the bees' access to the upper chamber. It was discovered that the queen rarely entered the upper chamber and therefore very few eggs were laid in there. The bees only used the upper chamber as a store for honey. In order to harvest the honey and wax the upper chamber was removed and the bees driven out. These bees returned to the lower skep and a second small 'super' skep was placed on top of the main skep to continue the process. Other designs were tried and became quite complicated.

Historical hive designs

The **collateral hive** had a square box for the brood with boxes on each side used to store honey. There were small channels between the central box and the side chambers to allow workers to pass from one to another.

In 1792 Francois Huber developed **a 'leaf' hive**. This hive was specifically developed to allow observation of the bees rather than for practical beekeeping. If the spacing between the 'leaves' was just right the hive could be opened like a book to observe the activities of the bees in the nest.

The **Stewarton hive**, designed by Robert Kerr in 1819 was a particularly notable design. It was octagonal and comprised a series of boxes that stood on top of each other. Each box had horizontal slats from which bees would hang comb. The gaps between the slats could be opened and closed to allow the bees access to the upper chambers.

In 1834 Major Munn developed his **Munn hive** that had removable frames. He discovered that if the spacing between the frames was a specific distance of between 6mm and 8mm they did not get stuck together with the wax comb built by the bees.

In 1851 the **Reverend Langstroth** patented a design of hive that had movable frames with a spacing of a quarter inch to ensure that frames could not be stuck together by the bees. In effect he promoted the concept of bee space. This is the gap that bees will leave between frames Although this development occurred over 150 years ago it was not until the late 1970s that the common use of skeps for holding bees finally fell out of use in Europe

Modern frame hives

Modern removable frame hives all preserve the need to retain bee space between every component within the hive but the internal designs can vary by changing the size and shape of the frames for the convenience of the beekeeper.

In 1935 a **British Standard for a beehive**, initiated by the British Beekeepers' Association was published. This standard has been modified over time. One of the most popular hives in the UK is now the Modified National (see page 41). Other hives have been developed either for greater simplicity of design and construction or to increase the volume of the brood chamber. However, most UK hives will use the same super boxes based on the same British Standard.

As well as refining hives used for European beekeeping, work was done to develop a hive that would meet the needs of beekeeping in Africa and to find a better way to keep bees than in the traditional log hives. Movable frame hives were considered inappropriate because of the precision needed to make the components and the cost of production.

In 1965 J.D. Tredwell and P. Patterson designed the Top Bar hive as a simple, cheap but effective box in which to keep bees. This was further developed in 1976 and became known as the Kenya Top Bar (KTB) hive and is still popular in Africa today.

Frame sizes

After reading so far you may think that there are only a few choices to make with frames.

Dig deeper and you will be surprised to find that there are over 100 different frame sizes and styles that have evolved over the years, from modern all plastic frames with foundation to home made varieties.

It's important to ensure that you are using the correct size for your brood box and supers.

Top: A selection of similar sized frames.
Bottom: Top Bar hives in Africa.

Bee calming

As an alternative to smoke, water sprays are effective as a means of calming bees.

A light spray of water tends to reduce the bees' ability to fly and may possibly confuse them sufficiently for the beekeeper to be able to observe the contents of the brood chamber.

Co-incidentally a fine dusting of icing sugar is also used in some Varroa treatments.

Above: A modern stainless steel smoker with a wire cage to help prevent burning.

Many natural materials are used for providing a cool smoke – wood shavings, dry grass, dry leaves, rolled corrugated cardboard (provided that it has not been fire retardent treated) and dry decaying wood as seen in the picture at right.

This latest development has recently seen favour with some European beekeepers; 'natural' beekeepers believe that this is more like a natural cavity a swarm would choose when establishing a new colony. There is no evidence that honey bees are at all concerned about the design of modern frame hives and would 'choose' a Top Bar hive in preference. While the Top Bar hive has advantages in terms of cost and simplicity, it can be more difficult to manage the bees in the hive.

Smokers

One of **the most important pieces of beekeeping equipment** is the smoker. This, like many other elements of beekeeping, has been developed over the ages. The action of smoke on bees seems to be quite complex. It has been shown that when smoke is puffed into a hive the bees stop guarding the nest and may well gorge themselves on the honey stores in the hive. This may be a natural reaction to the fear that there is a fire nearby and the colony may need to leave the hive. Bees communicate with each other by using pheromones (scents produced by one bee that has an effect on another bee). When smoke is wafted into the entrance of hive the acrid smell confuses the sense of smell that bees have and reduces communication in the hive.

The outcome is that the bees become disorientated and are in no condition to defend the colony. Smoke has also been used in the past to narcotise or kill bees (to access the honey stores). Tobacco smoke will narcotise the bees but when they recover they can often be very aggressive.

Honey hunters light a fire near to the nest and waft smoke into the nest before trying to take the honey. Plant material was usually used as this produced a lot of cool smoke.

The **earliest smokers** appeared to be clay pots with holes in the side. The beekeeper filled the pot with smouldering material and blew on the pot to make the smoke drift into the nest. This type of design is still used in some parts of North Africa.

From 1700 specially made clay pipes for beekeepers were produced and tobacco started to be a popular choice for smokers. Even 20 years ago some beekeepers in the UK were still using their pipes or cigars as a means of controlling bees.

The **modern type of smoker** was developed in America between 1873 and 1877 by Moses Quinby and modified by T.F. Bingham. The concept was that there should be a chamber for holding the smoking material along with a set of bellows that allowed air to be puffed through the smoking material and expelled smoke through a nozzle at the top of the chamber. In essence the design has not altered much over the last 100 years or so.

A gentle puff of smoke into the entrance before the inspection begins.

The beekeeping year

This section contains a brief description of the normal activities required to look after a colony of bees throughout a year. If the colony is managed correctly it will survive through the seasons and provide enough honey for its own use along with a surplus for the beekeeper.

Near the end of the 20th century, it was possible to allow a colony of bees to look after themselves through the year. They were able to swarm and some might die, but because there were many feral colonies living in the wild without any intervention from the beekeeper, there was no danger that honey bees would become scarce and their numbers insufficient to conduct their essential pollination activities. Today honey bee colonies need help to survive and, because many colonies are kept in urban areas and swarms can become a nuisance, it is important that the beekeeper looks after the colonies in his care. During the year some things will go wrong and the bees will not do what they are expected to do. As the beekeeper learns more about the life of a colony it becomes easier to 'read the bees' and understand how to help them thrive. The starting point is to have a routine of management that ensures the bees do not become a nuisance and helps the colony to stay strong and healthy.

Honey bees, like all animals, are affected by the environment they live in. In a temperate climate bees need to survive times when food is not available because of cold winters, dry seasons when flowers do not produce nectar or pollen, or simply periods when the weather is so wet that flying is not possible. Honey bee colonies follow a general cycle as the seasons come and go, but this cannot be aligned exactly with specific dates. Global warming and year-on-year seasonal variations mean that nothing is certain. Recent years have seen abnormal weather conditions both in summer and winter.

The management of a honey bee colony throughout the year cannot be precise because of seasonal and geographic variations and it is always a good idea to take advice from local beekeeping friends. Nevertheless, this section will use months to give an approximation of the times when certain jobs should be done. Bees are dependent on flowering plants and their cycle of activity is more aligned to the timing of particular flowers than it is to specific dates. Research has been carried out over the years to observe how this works and there is now evidence that the flowering times of specific plants are a good indicator of appropriate times to monitor the progress of the colony.

Inspecting the colony

When starting out as a new beekeeper, it will take much longer to inspect a colony. This is primarily because of the wonder of watching bees in their own 'home' but also because it will take more time to assess the health and strength of a colony and to decide if there is any need to treat or manipulate them.

For a novice beekeeper a general inspection should take no more than 30 minutes but this will reduce to about ten minutes with practice and experience. In spring and summer it may be necessary to inspect the colony once a week, although there are ways to extend this to once a fortnight. In late autumn and winter there is very little to do to the bees apart from protecting them

Monitoring indicators

Some examples are:

■ *When the flowering currant (Ribes) is in bloom it is time to conduct the first disease inspection of the season and ensure the colony is building well.*

■ *When lupins (Lupinus) are in flower, honey bee colonies are most likely to swarm.*

■ *When the blackberry (Rubus) finishes flowering honey bee colonies will start reducing in size in preparation for winter.*

Record cards

from bad weather, from other animals that may try to steal their stores and lastly, ensuring that they do not run out of food.

In answer to the question 'how much time', each colony may require thirty minutes a week from April to September and maybe thirty minutes a month outside this period.

Colony numbers

As the year moves from January to July, the number of bees in the colony increases rapidly from about 10,000 workers to more than 60,000 workers with some drones. From July to December the numbers in the colony reduce down to 10,000 as the colony prepares to survive winter. The graph below shows the numbers of bees in the colony (in red) from month to month and also the number of eggs, larvae and pupae (brood) in the nest (in dark blue). The graph shows that the quantity of brood rises rapidly in early spring to ensure that there are plenty of adult workers in summer to collect nectar to convert into honey. The colony, in effect, anticipates the summer when the availability of nectar is at its maximum.

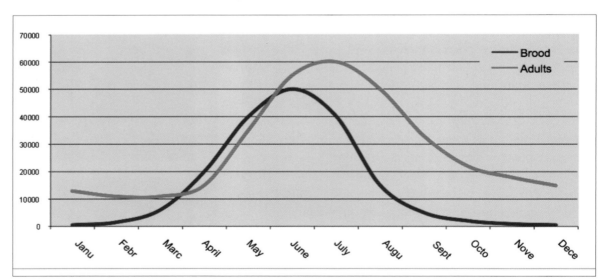

Of course if something happens to the colony, such as swarming or suffering an outbreak of disease, during the year the numbers will not follow this normal scenario.

Keeping records

Even with only one hive, it is sometimes difficult to remember
the state of the colony at the last inspection and to remember the
specific tasks necessary at the next inspection. It is a good idea
to get into the habit of taking notes on the current state of the
colony and the actions that should be considered next time the
colony is opened. Many different record card forms are available
that can be copied; alternatively many beekeepers develop
their own.

When starting beekeeping most beekeepers collect extensive
information from the colony to help them understand what is
going on. Here is a good checklist on indicators:

1.	Colony name	a unique reference
2.	Date and time	xx/xx/xxxx
3.	Weather conditions e,g. temperature	sunny/cloudy, rain, wind
4.	Temperament of the colony	1-5 where 1 = docile and 5 = very defensive
5.	Queen seen?	Yes/No
6.	Queen cells seen?	Yes/No including number
7.	No. of frames of brood	No. with eggs, larvae or sealed brood
8.	No. of frames of bees	No. more than 30% covered
9.	Quantity of stores in the hive	Rough estimate in kg or lb
10.	Space available for the colony to expand	Yes/No
11.	No. of frames in brood area clear of brood	No of frames
12.	State of the frames	Any old or broken needing replacements
13.	State of the hive	Any repairs needed to keep bee tight
14.	No. of supers on hive	To keep record of potential honey crop
15.	Action taken during visit	What you did to the colony
16.	Action required next visit	What you need to do next time

This is a very comprehensive list but as time passes and the
beekeeper gets more experienced many of these items can be
removed. After some experience a beekeeper will know if a
queen is present without seeing her, likewise if queen cells are
seen an experienced beekeeper will take action, depending on
the circumstances. Items 7 to 13 are dealt with at the time or
included in item 15. Similarly item 3 can affect the temperament
of the colony and is taken into account when assessing the
temperament. An experienced beekeeper will know how to deal

Medication records

Medication records are now a legal requirement in the UK.

The BBKA can supply a pro-forma card from their website or you can keep your records in a suitable note book.

Whatever you choose, you must keep the records for at least five years.

Inserting Bayvarol® strips into the hive brood box.

with a highly defensive colony and will just record the action. The authors know of very few beekeepers who keep no records at all. Some who keep no written records use the hive roof and symbols to record the state of the hive. This form of 'shorthand' works well for beekeepers who have kept bees for many years and need only minimal information about the colony history before the next inspection. Usually beekeepers who have reached this level will still have a notebook to use as a reminder about what equipment might be needed on the next visit.

There are a number of pro-forma record cards on the internet and many beginners' courses provide a template that will help beekeepers when they begin. The advice is to develop an approach that one finds suits the purpose and does not get constrained by a standardised form.

It is now a requirement for all beekeepers to maintain a record of all medication given to the bees. The BBKA has a sample record card on their web site that can be used. These days there are many substances that can be given to bees to improve their health or resistance to disease. Some of these substances have gone through rigorous trials and have been licenced as 'bee medicines'. In all cases use of these substances must be recorded with details of the date, quantities and batch numbers. There are other substances that beekeepers add to colonies of bees that either are not considered as medicine (i.e. icing sugar) or have not gone through the licencing process (i.e. certain organic acids). In these cases it is advisable to keep records of when and how much was administered as there could be an issue later on. Certain conditions (like Varroa) will always be a problem for honey bees and it is probable that new medicines and control substances will come on the market to help reduce the threats these conditions present to bees. Always record what is put into a bee colony. It may never be a problem but should an issue occur in the future the records may help to reduce any problems.

Take the record card with you whenever you plan to inspect your colony, as it will help you to remember to look for the most important indicators. Some beekeepers keep their record cards under the roof of the hive wrapped in a plastic folder. It is not advisable to leave paper inside the hive: the bees will chew it up and remove it from the hive. Bees seem not to attack plastic folders and will leave the record card alone if it is securely wrapped. Record cards should be filled in using a pencil or waterproof ink to ensure the information is not lost! In general there should be one record card for each hive and it is a good idea to start a new record card for each year.

The start of the beekeeping year

Because a honey bee colony can live for many years and carry on through winter with significant numbers, the start of the beekeeping year does not necessarily fall on 1 January. Most beekeepers prefer to start their year in September or when the honey crop has been removed from the hive. This is the time for winter preparations and if the bees and their beekeeper do not prepare adequately the colony could well die before next spring.

Preparing for winter

Once any honey has been removed from the hive it is time to ensure that the colony is in good health and ready for winter.

Wasps and robbers

In early autumn, most **wasps will be dying** of starvation and will seek out any source of sweetness and protein. Hungry wasps will raid bees' nests for the provisions within and will kill many bees if left to their own devices. The hive needs to be protected and the best way to achieve this is to ensure that there is only one entrance and that this is as small as possible. Ensure that there are no holes between components of the hive where wasps could gain access particularly through ventilation channels in the roof and that it fits well with no gaps. With a small colony, the entrance can be reduced so that it can be guarded by just one bee, but with a large colony the entrance should be reduced to about 70mm wide and 8mm deep.

Most manufacturers of beehives supply **an entrance block**, a piece of wood that goes into the entrance to leave just a small gap. Open mesh floors should be inspected to ensure there are no gaps where wasps could get into the hive.

When to buy bees?

Do not buy a colony of bees in the autumn. If a bee colony is not healthy and well prepared for winter it could die.

It is far better to buy a colony in the spring or early summer whilst the colony is expanding.

Let the seller nurse the bees through winter — it will show how well prepared they are.

Wasps have found a small gap under a feeder and are taking advantage of the syrup.

Wasps can also be diverted from the hive by providing bait traps near the hive. These must not be charged with honey: provide sweetness in the form of sugar or jam instead. If honey is used it will attract bees as well as wasps and the colony will be weakened.

Varroa treatment

Varroa (see Chapter 8) is **a major problem for honey bees** worldwide and all colonies will to a greater or lesser extent contain varroa mites. If the number of mites exceeds the maximum recommended by APHA (the Animal and Plant Health Agency – an executive agency of DEFRA) the colony needs to be treated to reduce the numbers of mites. At this time, medications containing thymol (oil of thyme) as the active ingredient can be used. Thymol kills mites but can taint honey, so it should not be used when honey for human consumption is in the hive. It can also kill bees, or severely disrupt the organisation of the hive, so should only be used according to the manufacturers' instructions. When applied correctly and with the right weather conditions (continuous daytime temperatures above 15°C) this treatment will destroy 80 to 90 per cent of the mites in the colony.

Adult bee and brood diseases

Early autumn is a good time to **check the health of the brood** and the adult bees. There are a number of brood diseases that can affect the colony and some of these are statutorily notifiable to APHA. Fortunately, the number of notifiable disease incidents identified in a year are currently few and most beekeepers go through their beekeeping careers without encountering them. Nevertheless it is important to check that they are not present and this is covered further in Chapter 8. If in doubt about any diseases, it is always better to seek advice from a beekeeping colleague or a government-appointed bee inspector, than to leave the colony untreated and hope that it will not be a problem. Beekeepers and government inspectors are usually friendly and helpful and will not criticise you or charge you if the call is made in good faith.

Brood diseases may also be present and again it is important to be able to recognise the look of healthy brood and seek advice if

A container of thymol (Apiguard®) being placed above the brood frames.

Collecting a sample of bees ready to be checked by the RBI or sent to the NBU for disease checks.

there are any signs of abnormality. It is difficult to be certain which diseases you are dealing with when starting beekeeping but with experience most can be recognised.

Adult bees also suffer from diseases and this is a good time to check if they are present. It is quite difficult to identify a sick adult but taking a sample of 30 bees and having them analysed for disease by an experienced beekeeper with a microscope will identify the most debilitating that can be treated (Chapter 8 has details of Nosema and this procedure). Nosema is a disease that affects the gut of adult bees and can reduce the vigour of the whole colony.

Winter bees

Early Autumn is the **time to check** that the queen is laying well and that there is plenty of room for her to raise new brood. The eggs laid now should result in adult bees that can survive the winter months. To live six months and take the colony through winter, these larvae must be well-nourished and free from disease. The queen should be present and the brood pattern should be even. There should also be many adult bees in the colony.

The other important task at this time is to check that the colony has sufficient stores of honey to last through the winter. To be sure, the colony should have about 20kg of honey for the bees to live on. To give an idea of what this means, 20kg of honey is equivalent to five full frames in a Commercial, Deep National or Jumbo Langstroth hive. With a National hive the colony will need a super full of honey and three frames in the brood chamber also full of honey.

The beekeeper needs to **estimate how much additional honey** will come from natural sources such as ivy, Himalayan Balsam or Rosebay Willowherb flowers by determining how abundant these plants are near the colony and by talking to local beekeepers. If the colony is short of stores and unlikely to be able to gather more in the next month or so, the colony supplies need to be supplemented with a sugar solution. Now is the time to feed the bees with a proprietary sugar feed or with white caster sugar dissolved in warm water. The sugar solution should be as

Opening a hive

Always be gentle when opening a colony, for quick movements and loud banging noises irritate bees.

Use enough smoke to calm the bees, but don't overdo it. Too much smoke and they become overdosed and soon may not react to smoke at all.

If your bees are not very gentle and become a nuisance you should requeen the colony.

The scourge of river keepers and many anglers is the Himalayan Balsam (*Impatiens grandiflora*). Although a non-native, and an invasive species, it is a useful plant for bees looking to increase their winter food stocks in late summer and early autumn. The plant has spread widely and so far has proved to be almost impossible to eradicate from many waterways.

Protecting hives

These hives have been protected with plastic sheeting to prevent Green Woodpecker damage.

In addition mouse guards are fitted to the entrance of both the near hives. Around the farthest hive there is a wire mesh tied to an old tree stump and supported by metal stakes – again to prevent woodpecker damage and any possible incursion into the apiary from badgers.

The red 'card playing' symbols on these hives (right) are said to help bees find the correct hive when returning after a flight away from their home.

concentrated as possible. The solution in placed into a special feeder that is put on top of the brood chamber underneath the roof. The bees will then come up inside the hive to collect the sugar solution, process it into honey and store it in the hive.

Protecting the hive

In October it is time to ensure **the hive is still protected** against other animals that may try to get access to the honey stores or the bees. Again ensure there are no entrances except the main one used by the bees. In some areas Green Woodpeckers (*Picus viridis*) have learned that hives are good sources of food. Where this is a possibility the hive must be protected. Woodpeckers will peck through the wall of the wooden hive so that they can use their long tongues to feed on honey, larvae and even bees.

A rapid feeder containing syrup as explained in the text which is placed on the brood box.

Mice may try to gain access to the warm interior so that they can spend the winter in relative comfort. The entrance must be made small enough to stop mice entering the hive. If the entrance is less than 7mm deep the mice will be unable to get through it. Alternatively, proprietary mouse guards can be bought from beekeeping suppliers and pinned in front of the hive entrance. Finally badgers and other large animals can knock over or dislodge the hive and, if this is a potential problem in your area, all the hive components should be strapped together and pegged to the ground.

Now it is time to relax and enjoy the winter months in the knowledge that your bees are safe and have plenty of food to last the cold months.

Second Varroa treatment

At the turn of the year the queen will **probably have stopped laying** or reduced her laying rate to a minimum. This is a good time to treat with an oxalic acid medication to further control Varroa mites.

You should only treat at this time if the number of Varroa mites remaining in the colony following the autumn treatment remains substantial (details in Chapter 8). The treatment is better done on a warm sunny day when bees can be seen flying from the hive. This will mean that the cluster of bees has loosened and the other bees are mobile in the hive. The oxalic acid medicine should be dribbled over the bees as quickly as possible and the hive reassembled with minimal disruption.

The old year's **work is now complete** and as long as the tasks have been completed successfully the colony should be safe and in good shape to await the start of spring.

Once the external temperature goes below about 14°C the worker bees cluster together to conserve body heat. The cluster surrounds the brood in the nest, which must be kept at about 34°C for optimum development. The bees on the outer edges of the cluster must not allow their body temperature to fall below about 7°C otherwise they become torpid and could drop off the cluster to die. The cluster is therefore always in motion with the outer bees moving inwards and their places being taken by bees that have been in the middle of the cluster. Using this strategy, colonies of bees can withstand external temperatures far lower than ones experienced in the UK so long as the hive is kept dry and reasonably ventilated.

As the external temperature moves nearer to 0°C for extended periods the cluster will tighten to reduce heat losses but eventually the already reduced laying rate of the queen will stop altogether. After three weeks there will be no brood in the nest and, at this stage the temperature in the middle of the cluster will

Treatments

To successfully manage honey bees you need to be able to identify what pests and diseases they are likely to encounter and know what options you have to control them.

You also need to know what the pest population is and at what stage you need to take action now and in the future if no control measures are taken.

It is inadvisable to use home-made treatments because of the dangers of contaminating honey or killing bees by using the wrong dose.

Oxalic acid being used to treat the brood.

Check food stores

Early spring – any time between late January and mid March depending on the weather and where you live, is an important time for bees as they are most probably already raising brood and could be running out of stored food supplies.

Double-check the food supplies and if you are in any doubt add fondant – better to be safe than sorry and lose your colony due to starvation. If they take the food down you will quickly know whether they are short of stores.

Hefting the hive to check that there are sufficient food stores for the bees' winter survival.

drop to about 20ºC. This conserves energy coming from the food stores and means the colony can survive longer. With sufficient stores a colony of bees scan survive many months in very cold weather and yet be ready, once the temperatures rise, to continue their development.

Spring

As spring arrives **the queen's egg laying will increase** and the colony will now need to keep the brood nest at a constant 34-35°C to ensure the brood develops without damage. To keep the temperature high the workers will consume large amounts of honey and there is a danger at this time that the colony may run out of food.

Feeding

At fortnightly intervals **the colony needs to be checked** to ensure that there is sufficient food in the hive. This is done by 'hefting', a procedure that does not require the hive to be opened. The beekeeper lifts one side of the hive just off the ground in order to feel its weight. It takes some practice to accurately judge the weight but if the hive feels light and easy to lift, it is very short of food for the bees. It should feel heavy and be difficult to lift. Hefting can be practised with empty hives loaded with 20kg of any material such as stones or earth (in a bag of course!).

If the hive feels light (say less than 10kg), then the bees must be fed as soon as possible. The easiest way to feed them in early spring is to place a bag of fondant, specifically formulated and obtainable through beekeeping suppliers, or bakers' fondant (the icing on iced buns and available from bakers shops), on top of the brood frames where the bees will come and collect as much as they need. Some beekeepers will place a bag of fondant on the top of the brood chamber throughout winter so that it is always available to the bees if they run out of other sources of food in the hive.

Colony losses

In March the 'winter' bees that **kept the colony** alive through winter will be dying, having completed their work. If for any reason they were not in top health they may not survive the required six months and the reduced number of winter bees may not be able to keep the colony alive until the new eggs laid in January and February have developed to take over the duties of maintaining the queen and colony.

Colonies may **die at this time** because the winter bees were not sufficiently fit. It is the beekeeper's responsibility to try to ensure this does not happen. Despite every effort some colonies will die and, before the arrival of varroa the annual average number to die during the winter months was about five per cent. This number has increased rapidly since then and in the UK in 2007 it was about 30 per cent (in the USA it reached more than 40 per cent!). By 2010 the UK figure had reduced to about 12 per cent, which is still too high. The reduction was in part due to a series of hard winters at this time.

Although it may seem strange, colonies of **bees survive better** and use less food if the winter temperature stays between +5°C and –18°C. At higher temperatures the cluster of bees breaks up and the activity consumes more stores and below this temperature range the bees need to work harder to keep their temperature above +7°C.

First inspection

Mid-March is the usual time to have a quick look inside the hive to ensure that all is well and the colony is building up with plenty of bees and brood. Choose a warm (above 14°C) and sunny day when there is not too much wind. Wear your bee suit and do not take chances. It can be difficult to assess the temper of a colony after it has been left alone for a couple of months. Before opening the hive it is worth just watching the entrance for a few minutes.

If all is well then there should be:

- Plenty of bees flying in and out of the hive
- Some bees returning with their pollen baskets full of pollen (which is usually yellowy green at this time of year)
- A few dead bees below the entrance.

If the bees are collecting pollen and bringing it back to the hive

Winter colonies

The BBKA conducts regular, and completely random winter bee surveys amongst its members to establish winter colony strengths and any winter loss trends.

Feeding well in the autumn to ensure adequate honey stores and a Christmas present to your bees of a block of fondant will help ensure their survival until the first warm days of spring.

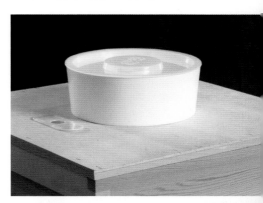

Top: A small capacity liquid feeder for emergency use if a hive is short of food.

Bottom: A bag of specially prepared fondant – a quick and easy food source in cold weather – cut open and placed on top of the brood frames.

First inspection

First inspections are always tricky to time, especially if you are working and only have weekends available to look at your bees.

Remember to watch the weather and try to anticipate a likely day when most variables are going to be right for you and the bees.

Go through your mental check list of all the things you will need, ensure that everything is clean and you have properly zipped up your beesuit. Do you have a pair of gloves?

Take a few minutes – maybe 15 or so to see what the bees are doing – then go for it.

this is a good sign that the queen is laying eggs and brood is developing. The bees prefer fresh pollen to stored pollen and will collect it to feed the larvae whenever it is available. However, if there are no larvae in the hive for whatever reason, the bees will not bother to collect pollen.

The dead bees are there because during winter **some individuals will die** and when it is really cold the corpses are simply left where they fall. Once the temperature warms up the workers will tidy the hive and remove any dead bees and drop them just

outside the entrance (it will still be too cold to carry the bodies any distance from the hive). There can be as many as 200 bees outside the hive at various stages of decomposition. If however there are thousands of bees on the ground then something dramatic has happened and the colony will be at risk.

After checking the outside of the hive it is time to look inside. The key is to try to do this quickly without disturbing the bees any more than necessary.

Disassemble the hive to expose the brood chamber. At this time of year the outer frames are usually free of bees and certainly free of brood and can be removed to give space for inspecting the remaining frames. Remove each frame so that the contents of the brood chamber can be assessed. The hive is now ready for a quick inspection and it should not have taken more than two minutes from taking the roof off the hive to get to this stage.

Centre: A returning worker bee with a full load of yellow pollen.

Above: a quick spring inspection during a beginners day at the apiary.

Inspect each frame quickly to examine the brood and the stores. In this first inspection, the purpose is to determine whether the queen is laying well and there are plenty of stores to keep the colony going. There are many other things to look at and bees can be fascinating, but this is not the time to enjoy your beekeeping at the expense of the bees who will get chilled and try to defend the nest.

A quick winter inspection to see how the bees are doing will be followed by an early spring visit to check on food stores and general colony wellbeing.

This should take no more than ten minutes at this time of year so that not too much heat is lost and the brood becomes too cold. Once the hive is returned to its normal state the bees will cover the brood to bring it up to the normal temperature and clear up any mess that you have left.

Is the queen present?

It is always a good idea to ensure that **the queen has a paint mark** on her thorax (see page 113 for more details on how this is performed) so she can be seen easily. Sometimes, over the autumn period, the old queen may be superseded by a new queen with no identifying marks. In practice, at this time of year it is not necessary to see the queen, only to know that she is in the hive and laying eggs. Every beekeeper should be able to see eggs in the cells but this can take some time until the eye is practised. The other way to know if a queen is present is to look for larvae and sealed brood. If there are larvae present then a queen must have been in the hive about a week ago (an egg takes three

Good winter stores on a deep National hive frame.

days to eclose [hatch] and a reasonably-sized larva could be three days old).

Are there enough stores?

When you are going through the brood box it is a simple task to add up the weight of stores (honey) on each frame, remembering that at this time of year there should be at least 10kg of stores present. Depending on the size of the frames this can be calculated by adding up the parts of frames containing sealed honey. For a Commercial frame or Deep National frame each equivalent to a full frame equates to 4kg so two and a half frames will be sufficient. If the frames are National frames then four frames will be required.

Having made sure the colony is developing and has enough stores it is now better to wait until mid or the end of April before looking at the colony again.

Above: a marked queen makes it far easier for the beekeeper to spot her when carrying out regular hive inspections.

Centre: Bees storing away honey for further ripening. When the honey has reached the correct water content (around 18 per cent) it will be capped over – seen at top right – with new white wax.

First major inspection

This should be done **when the flowering currant** (*Ribes*) is in flower, usually in late April. If there is no flowering currant nearby then the appearance of dandelion flowers is a good indicator. The weather should be warm (hotter than 17°C) and sunny – the type of weather that is sometimes called shirt-sleeve weather. Now is the time to have a good look at the colony to ensure it is developing well and expanding in numbers.

The hive should be opened as usual to expose the brood frames. This time the frames should be looked at more carefully and it should be possible to find the queen, even if she is not marked. This is a good time to mark and possibly clip her (see page 114) so that she can be found more quickly later in the year when certain manipulations will be much easier if she is. At this time of the year there could be about 20,000 bees in the hive and it will be easier to find the queen now than later on when there could be 60,000 bees. The frames should be inspected to ensure that the brood looks healthy (see page 81) and old or damaged frames can be moved away from the centre of the hive to the edges.

It is important that the brood nest is not disturbed too much because if a frame of brood is separated from the rest, the bees may think that the queen is no longer there and this can encourage the colony to make preparations to swarm. If one of the frames with brood on it looks ready for replacement, place it towards the edge of the brood nest rather than the edge of the brood chamber.

Take a sample of **30 bees to test for Nosema** (see Chapter 8 for details on this procedure) and send this to a colleague who is able to analyse the bees for the disease.

The final job is to assess whether the colony is developing well and will be able to build up during spring and early summer. The size of the colony at this time of year will depend on:

- The weather.
- The quantity of forage that is available to the colony.
- The geographic location.
- The health of the colony.

The first three are outside the control of the beekeeper once the apiary has been established and can vary from year to year, so absolute numbers for the size of the colony can be misleading. However the brood should be on about five frames and there should be about 20,000 bees in the colony. This approximates to about six frames being well covered with bees. A better way to

A busy bee

An adult worker honey bee in good condition weighs in at roughly 80mg and will be able to carry on each of her foraging flights between 40 and 60mg of nectar.

It therefore takes a considerable amount of foraging effort and lots of flying time to produce a kilogram of honey.

Flowering Currant (*Ribes sanguineum*) in flower is a good indication of when to start the first inspections.

Look out for flowers

Dandelions and celandines can create a swathe of yellow when in full flower and both are good indicators of when to make your first major colony inspections.

As the weather warms, you'll notice these bright yellow flowers along most roadside verges and fields from March.

Dandelions (*Taraxacum officinale*) appear as the days lengthen and remain warm.

know if the bee numbers are building as expected is to ask colleague beekeepers in the same area how their bees are doing and, bearing in mind a bit of exaggeration, this will give you a benchmark.

Finally, **check the approximate ratio of eggs** in the colony compared with sealed brood. If a colony is just about holding its own then there will be four times the number of cells with

sealed brood compared with eggs (the period eggs last is three days whereas sealed brood lasts for 12 days). If the proportion of eggs is greater than this, the colony is doing well and expanding. If the proportion is less, the colony is weak and reducing in size. Once the inspection has been completed, reassemble the hive and update your records for the colony. After this first major inspection the frequency with which you should inspect the hive increases.

May/June inspections
During May inspections should be weekly and the aim is to monitor the progress of the hive and to watch for swarm preparations.

Every time the colony is inspected, ensure that the five key aspects of inspection (page 96) are followed so that the state of the hive can be assessed.

When inspecting the brood frames **look carefully on the frames** for signs of queen cups or queen cells (see Chapter 7 for more details). If these are present, look inside them to see if any have eggs or larvae in them. If the inside looks shiny and polished then the colony is preparing to swarm. Once the colony has decided to produce a new queen it is difficult to stop and the best action is to perform a method of artificial swarm control. This is described in more detail in Chapter 7. Not all colonies will try to swarm every year; it depends on the age, health and egg laying capability of the queen, the space in the hive available to the queen to lay eggs and space to hold the number of adult bees without congestion.

Once the colony has been inspected and it is clear that it is still building and not trying to swarm it is time to reassemble the colony. However, before this, dust the frames with bees on with icing sugar to reduce the number of Varroa mites in the hive (see page 165 for how this is done). This is only effective if the hive has an open mesh floor.

When the hive is reassembled, a super can be placed above the queen excluder. If the hive has been over wintered without a queen excluder and super then this is the time to replace them. The super will give the bees more space and reduce congestion in the hive. It will also give the bees space to process nectar and to store honey. This will again increase space in the brood chamber and reduce the pressure on the bees to initiate for swarming.

At each weekly inspection consider adding another super. Unless there is a major source of nectar nearby (such as oil seed rape) it should not be necessary to add more than two supers before the main nectar flow.

In May and June additional supers should be added when the previous super is just about half full of honey and nectar. There is controversy about how to add supers. If the frames in the super are fitted with foundation and have not been drawn out into honeycomb it is better to place the new super just above the queen excluder with any other supers above the new super. This is the warmest place in the hive and will help the bees to raise

Inspections

Keep good records and always refer to them before carrying out any inspections.

Have a plan when opening a colony, know what you want to do and stick to it.

Become familiar with how a honey bee colony works and what you should be finding each time you open up the hive. Check for a healthy colony and for signs of disease or any pests.

Are there plenty of honey and pollen stores and a good mix of workers and drones?

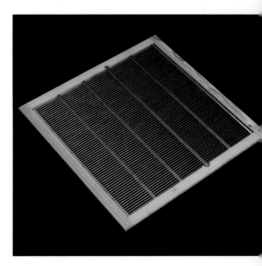

A wire rod queen excluder. This style is used by many beekeepers as the rods do less damage to bees' wings as they pass between the rods than the perforated type of excluder.

their temperature to start to produce wax. Placing supers in this position rather than on top of the existing supers also encourages the bees to move over the frames and to start to use the new space immediately.

With supers containing frames with drawn honeycomb, it is possible to place the new supers above the existing ones. This is easier as it does not require additional lifting but may not encourage the bees to move onto the new super.

If the weather is warm and sunny there may be an abundance of flowering plants and the bees will be able to store a lot of honey. Under these circumstances it may be possible to remove some honey for your own use. However there can be periods when the weather is poor and the bees may not be able to find food. After a few weeks of bad weather the remaining stores in the hive may be consumed and the colony could starve. If at all possible leave the honey for the bees. If you do not have enough supers then some honey must be removed from the hive and the empty super replaced.

July inspections

Depending on the forage available in the locality July can often be the main time for bees to collect nectar and build up their honey stocks. By this time the likelihood of swarming will have reduced and the beekeeper's role will be to keep the colony together and as strong as possible. While bees are away from the colony foraging it will not be difficult to inspect but there could be as many as 60,000 bees in the colony at this time of the year.

The rate at which supers are added should be increased during July to ensure that the bees have plenty of space to process the nectar into honey and then seal the honey with wax to ensure it remains stable and will not absorb water from the atmosphere.

Inspections later in summer can be relatively quick, but it is still important to look for queen cells otherwise the colony may swarm and half of the workers will be lost and the colony will not be strong enough to bring in a surplus of honey.

Queen cups built along the bottom bars of a brood frame

At the end of July or early August the main nectar flow will reduce considerably and the larger colonies will become very defensive. They will be attacked by wasps and by other colonies of bees that will be trying to rob the honey from the colony. Wasps are a major problem and once they find a weak colony they will be relentless in their attacks, killing the bees and trying to get inside the hive. Weaker colonies will often succumb to these attacks and can be destroyed if they are not protected by the beekeeper (prevention measures are described on page 151).

The end of the season

The end of the season is **marked by the dramatic reduction** in the availability of pollen and nectar in the locality. In general this is in August or September, but can be earlier or later by at least a month depending on where you live and the weather.

At this time the beekeeper's main job is to **remove the surplus honey** from the colony and ensure that the bees are able to defend themselves from any attacks. The removal of the honey supers is described in Chapter 9. This needs to be done quickly without disturbing the bees too much. At this time the bees will be in the mood to attack any predators (including the beekeeper) to stop them stealing the honey.

If there is more than one hive in the apiary, the removal of supers is best done in the evening, when fewer bees are flying. The process of removal exposes a large area of the hive and the aroma of honey will attract many other bees and wasps looking for a free feed.

Once the **supers have been removed** and the honey extracted from them they can be replaced on the hive (above the crown board but below the roof). This will encourage the bees to come to the supers and remove any honey left on them by the extraction process. After about three days the supers will be cleaned and 'dry'. They can then be removed from the hive and prepared for storage until the next year.

The season is over and it is now time to prepare for the following season and consider how to be a better beekeeper in the

Brood boxes and supers, together with frames that have had the liquid honey extracted, are now stacked ready for cleaning by the bees.

Hive tools

Purchase good quality stainless steel hive tools with a thin and tapering blade. You will find that good quality tools will last for many years.

Thin blades can be more easily pushed between the hive brood and super boxes to break any seal that the bees have made with propolis without damaging the hive components.

The beekeeper has removed the roof of the hive and placed it on the ground 'upside down'. He is now removing the super and crown board (cover board) and will place these items onto the waiting roof. He will then start his inspection of the brood box.

Note that he is holding the smoker between his knees – ready to smoke the bees again as he will have already given them a puff or two before he lifted the super.

future. Your local Association will probably be running winter courses to teach more advanced beekeeping techniques and this will help you to further your understanding of bees and beekeeping.

Important manipulations

1. Opening a hive

First give the **entrance a quick smoke**. The objective is not to flood the hive with smoke but to let it waft into the entrance to alert the bees that something is happening. Wait a couple of minutes for the bees to settle and then lift the roof off the hive and gently smoke across the top. The roof should always be placed upside down on the ground and all the internal parts placed on top of the roof so that they do not touch the ground and become contaminated.

Next **remove any supers** on the hive along with the queen excluder if it is there. This will expose the brood nest, which is the only part of the hive that needs close inspection.

Supers and other internal parts of the hive tend to get 'stuck' together by the bees using propolis or by bridging gaps with

wax comb. Insert the hive tool between the two components that you wish to separate and lever gently. This will help to break any seal without too much disturbance to the bees. Finally, when lifting a super it is advisable to rotate it slightly. This will help to break any links between the frames in the upper super and any in the lower super. If this is not done then it is possible to lift a frame out of the lower box with the inevitable consequence of disturbing the bees and possibly dropping the offending frame onto the ground.

Waft smoke across the top bars of the hive to encourage the bees to keep on the frames and not fly upwards.

With the hive tool, **move all the frames** to one end of the brood box. This is done by placing the hive tool between one end of the hive and the sidebars of the first frame and then levering.

Using the hive tool

When you start beekeeping, get into the habit of keeping your hive tool in your hand while you carry out regular manipulations.

You will find that doing so quickly becomes second nature and will save time hunting for a misplaced hive tool, or losing it in long grass!

The first frame has been removed and placed alongside the brood box. The beekeeper now goes methodically through the brood box moving frames into the gap left by the first frame, removing each frame in sequence to make his inspection.

The process is then carried out at the other side of the first frame so that all the frames are still parallel with the brood box but all tight and set away from one end.

Now the **dummy board or first frame** can be moved out of the hive using the hive tool and gently lifting it vertically. The frame is moved to the middle of the gap created and then slowly lifted

Queen marking

The international code for queen marking allows you to easily identify the year a queen was born, whether you bought her or marked her yourself.

Year ending in
>   *1 or 6 = white*
>   *2 or 7 = yellow*
>   *3 or 8 = red*
>   *4 or 9 = green*
>   *0 or 5 = blue*

Queens rarely live or are kept for more than five years, therefore the marking colours repeat every five years.

Above: a queen can at first be very difficult to spot amongst the mass of bees on a frame – particularly an unmarked one.

Most beekeepers therefore mark their queens with a dab of water-based colour. Here we see a 'princess' that has not been marked or had one of her wings clipped to help prevent future swarming.

Right: Beginners starting to inspect the brood.

so that any bees on it are not crushed or rolled. This frame then needs to be placed to one side, either in a nucleus box or against the side of the hive, but not touching the ground.

Some beekeepers use a specially made frame holder that clips on the side of the hive to hold the first one or two frames. This prevents the frames being damaged or touching the ground and becoming contaminated. Before placing it to one side, just check that the queen is not on it and that there is no brood on the frame. If there is, this frame should be placed in a nucleus hive to prevent the queen being lost or the brood becoming unnecessarily chilled.

2. Inspecting the brood

The process involves loosening a brood frame from the adjacent one and moving it into the space created by the removal of the first frame. Gently lift the frame out of the hive and inspect the

side facing you; once it has been checked, turn the frame over and look at the other side. If it is necessary to look carefully at the brood it might require the bees to be shaken off the frame. Hold the frame in both hands just inside the hive and shake with a sharp down/up movement. The bees will drop off the frame into the hive and the brood will be clear of bees so it can be clearly viewed.

When inspection of the frame is complete, place it carefully back into the hive and move it away from the next frame to be inspected. In this way the 'gap' in the hive moves across the hive until the last frame has been inspected. At this stage the gap will not be in its original place. Put the hive tool in between the hive and the first frame and gently lever it to move all the frames together and restore the gap to its original position. Then place the frame that was removed back in its original position. Centre the frames in the middle of the brood box and replace the queen excluder, any supers and the roof.

3. Finding the queen

There are a number of characteristics the queen possesses that will help the beekeeper to find her among many thousands of worker bees.

- The queen moves across the frames in a different way from the workers. She is more deliberate in choosing a direction and she will stop to inspect brood cells to determine if they are empty and have been prepared ready for her to lay a new egg.
- Her legs are longer than the workers' and she looks taller in the comb than others bees in the colony.
- Her abdomen is longer than a worker's and it appears that she is dragging her body across the comb.

Above: Finding the queen can be very difficult for beginners – especially for the first few times. Scan the frame in a spiral, starting at the top – you will quickly learn to spot the queen. You can also do a scan by adopting a zig-zag approach up and down the frame.

Don't forget to cover both sides of the frame and the outer frame edges as queens can be sneaky!

Top: Marking the queen makes life a great deal easier on each inspection as you will spot her more quickly.

Planning inspections

Remember to plan your apiary inspections before starting out.

- *What am I going to be doing on this visit?*
- *Have I the correct tools for the jobs I am planning to carry out.*
- *Do I have a clean bee suit, suitable footwear and gloves?*
- *Do I have a clean smoker, matches or lighter and plenty of fuel?*
- *Do I have a spare bucket with lid for old wax or other debris?*
- *Do I have a suitable container with cleaning materials for my hive tools?*

Make a list and run through the various items on it before setting off, particularly if you have a car trip to your apiary.

Where is she? There is an unmarked queen in this photograph. See how long it takes you to spot her.

The technique for finding the queen can be learned and relies on a sharp eye and patience. As each frame is removed from the hive it should be held at a comfortable distance so that the complete side can be seen easily. Quickly look around the edges of the frame, as the queen sometimes will walk round to the dark side. Once this area has been checked, scan the rest of the frame in a spiral or zigzag motion to cover the whole face of the frame. Sometimes the workers will be bunched up in part of the frame and it is worth placing the back of your hand over the bees so that they gently move away. If the queen was under the bunch she will be seen as she moves away.

The queen is usually found on a frame where she is laying eggs. When going through the hive, have a look for evidence of egg-laying and spend more time looking at this frame to increase the probability of finding her.

This technique should be adopted for both sides of the frame and then for all the frames in the brood chamber. If the queen is not seen, then the approach can be used a second time, but if the queen is not found then it is probably better to give up until another time.

Reading the bees

This term is often used to express an understanding about what is going on in the colony. Once able to read the bees a beekeeper can explain how the colony got to the current conditions and predict how it will fare over the period before the following inspection. You are acting as a detective and assembling clues. The process is best learnt from other beekeepers. Going to demonstrations and asking many questions of the demonstrator will help you to recognise the tell-tale signs that are used to assess the colony.

Here are few tips but these are best observed:

1. How populous is the colony?

The colony should build steadily in spring and early summer and then decline steadily towards autumn and winter. The number of workers in any hive will depend on the size of the brood chamber, the weather over the season and many factors like colony health, food stores and queen fecundity.

Comparing the colony size to an adjacent colony might indicate inherent disease or the possibility that the colony may have swarmed previously. A colony that seems very congested will be much worse in the evening when all the flying bees return. This alone could initiate swarming and without action bees and the queen could be lost

2. What is the demeanour of the workers?

Workers can become more defensive when there are stores in the hive that need to be protected. Conversely if there is plenty for the bees to forage, they will be too busy to worry about defending the colony. If the bees are not behaving as might be expected there could be an underlying disease problem. Varroa mites on the adult bees can lead to a more disturbed nature for the colony. If the bees seem strangely docile or excessively defensive look for underlying causes.

3. How active is the queen

During the active season, say from March to July, the queen should be walking over the frames detecting cells where she can lay an egg. She is not normally distracted from this purpose.

If she is acting lackadaisically this could be an indication that the colony is not finding sufficient food, or is preparing to swarm, or that the queen is failing and should be replaced.

Reading the bees

Experience is the best teacher when it comes to reading your bees and gentle bees are more fun to manage than feisty ones.

Even in a large colony you should not have many flying bees after a few minutes from opening the colony for an inspection. A light puff from your smoker will help to keep the bees inside, between the frames.

The bees should remain calm and relatively still on the top bars and when a frame is taken out to inspect should not fly off.

There should be no followers when you leave or feisty guard bees to greet you.

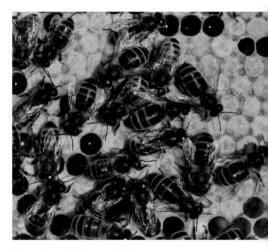

Remember that there will be a great many workers out foraging when you check your colonies over lunchtime, so you may not get a true picture of hive numbers.

4. Population ratios

Because eggs remain for about a quarter the time sealed brood remains, this can be used as an indicator of the general growth of the colony. If the ratio of brood to eggs is greater that 4:1 then the colony is not building but failing. This is fine in autumn but not during spring and summer. Similarly if the ratio indicates the colony is still growing in the autumn there is a danger that it will not have sufficient stores to survive winter. An exception may be a new queen in a small colony that is still building up for winter.

5. Floor debris

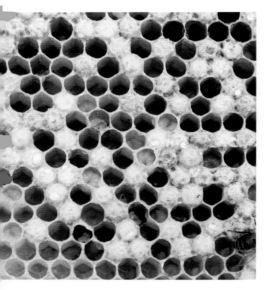

Above: Signs of a drone laying queen or a disease. Either way the colony should be carefully inspected for disease.

Right: Winter debris on the hive floor. You will find all sorts of debris here from spent wings to dead Varroa mites after a successful treatment.

The debris on the base of the hive is a good indicator of the general condition of the colony. Healthy colonies will keep the floor clean. Wax on the floor can indicate that the bees have started to use stored honey, little flakes of 'clear' wax can indicate that the bees are making new cells. Chalky lumps are an indication of Chalk Brood (see Chapter 8). Dead bees might suggest that the colony is being attacked and robbed.

6. Brood pattern

This usually indicates the sealed brood but can include open brood and eggs. The pattern should always be neat with few missed cells (the 'pepper pot' pattern). The brood capping should be a biscuit colour, slightly domed and uniform.

Variation from this norm can suggest disease or a failing queen. None of the above is an indicator on its own but should be taken with all the other 'clues' and what might be expected from this colony to assess its state. It may take a number of years of bee-keeping to be confident in analysing the state of the colony of bees but the real message here is that when inspecting a colony of bees it is not enough just to look and see if there is a queen and brood.

The key art of inspection is to keep an open mind and collect information on a wide range of factors and then to work out what all the information means. Once this approach is understood and followed it is possible to decide after each inspection what action should be taken to ensure the colony continues to evolve as it should.

Too many drones

A frame that has drone cells scattered all over it, can be the work of a drone-laying queen or it can be a colony that has laying workers.

In either case further inspection is required and prompt action taken as the colony is unlikely to survive as there will be insufficient worker brood developing to ensure its continuation.

Below: 'Pepper pot' brood pattern on a 14x12 frame, which indicates a failing queen or perhaps disease.

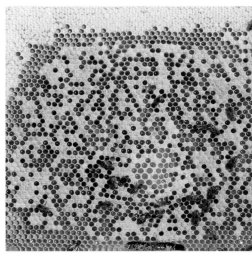

Left: Inspecting a National brood frame that has been specially placed in a 14x12 brood box to encourage the production of drone cells. Note the larger dome capped drone brood on the sacrificial comb at the base of the frame.

Top: A classic frame of sealed stores showing the semi-circular pattern of capped honey.

Bottom: A mature queen cell ready to be harvested and used in a weak colony that needs requeening. The beekeeper will have to carefully cut out a section of the frame under the cell to ensure that it is not damaged when cutting it out from the wax foundation before transfering to another colony.

7. Five key elements to inspection

There are many ways to inspect frames and to assess the state of the hive. In 1976, an excellent beekeeper and teacher, Ted Hooper, wrote a book *The Guide to Bees and Honey* in which he proposed five things that should be considered when inspecting a colony.

These are:
1.   Has the colony sufficient room?
2.   Is the queen present and laying the expected quantity of eggs?

Early in the season:
3a. Is the colony building up in size as fast as other colonies in the apiary?

Mid season:
3b. Are any queen cells present in the colony?
4.   Are there signs of disease or abnormality?
5.   Has the colony got sufficient stores to last until the next inspection?

With each 'question', if there is a problem, you must take corrective action to ensure that the colony is going to be in good condition next time you inspect it.

The actions are as follows:

1. If the colony does not have sufficient room then it is time to put on a super. If the congestion is in the brood chamber where there is just not enough room for the queen to lay eggs, it is important to give her more space by adding a new frame of foundation to the brood chamber and removing one that is either full of old pollen or old honey. Alternatively the brood chamber size can be increased by adding a second brood chamber complete with frames of foundation or drawn comb.

2. If the queen cannot be found and there is no evidence that a queen is there (no eggs are found) then see if the colony is making a new queen from a queen cell and, if not, add a frame of eggs and young larva from another colony of bees. You may have to ask your mentor for help if you only have a single colony.

3. If the colony does not seem to be building up well in the spring there may be a problem with disease. If there is no sign of brood disease then it may be that the adult bees are diseased or the colony was short of bees and/or short of stores through winter and is having difficulty recovering. Take a sample of 30 bees and ask an expert to check if the bees have Nosema (more detail on page 161), which can weaken the adult bees and reduce their lifespan.

Later in the season you need to look for signs of the colony preparing to swarm. If queen cells or polished queen cups are found, it is necessary to do something rather than just close the hive and hope for the best. It may be possible to remove all the queen cells, provide more space for the bees by adding a super and check again after a week. Normally once the colony starts to raise a new queen in the queen cells you must start a process of swarm control (Chapter 7) to ensure that you do not lose your queen and possibly half of the workers in a swarm.

4. A quick look at the brood pattern each time you inspect the colony will help to catch many diseases at an early stage. Disease can be the reason for the colony not building properly. Varroa can also delay the build up of the colony and you should check the number of mites in the colony (see page 163-4) and treat the colony if the number of mites is too high.

5. When looking through the colony make an estimate of the quantity of honey stored in the hive. A full super frame holds about 1.5kg honey and a full national brood frame holds about 2.5kg honey. To be safe there should be enough honey in the colony to last until the next inspection. Usually that will require at least 5kg honey in the hive during the summer season. If the stores are less than this the colony will require some feed.

Top: Two supersedure queen cells ready for transplanting into a new colony.

Bottom: Two supersedure cells – a sure sign that the old queen is failing. One of the cells has been opened up to show development of the immature queen.

Queens and drones

The queen is at the heart of any colony of bees. Unlike the other females (workers) who live brief, hardworking lives, she will live for a number of years inside the hive being fed and protected by the workers. As she ages her ability to lay fertilised eggs using sperm in her body will come to an end.

An unmarked and unclipped queen on a brood frame with honey stores. The unripe honey is clearly seen in the cells under and to the right of the queen's abdomen.

Before this, the pheromones she produces will reduce and her attractiveness to the colony will diminish until there comes a time that she must be replaced. The colony will start to produce new queens and will then either swarm to start another colony or sacrifice the old queen to allow a new queen to continue the work of providing eggs to make new bees.

Because the queen lays all the worker eggs in the colony, the bees are closely related to each other and her genetic makeup is key to the quality of the colony. A strong, healthy colony can only come from a queen with the right genetic characteristics. This chapter will describe in detail the life of the queen, the mating process and how a beekeeper can help to ensure that the colony remains strong and healthy by selecting the best queen to head the colony.

The life of the queen

A new adult queen will be ready to leave her specially constructed queen cell 16 days after the egg was laid by the previous queen. The wax casing round her cell is so strong that she cannot get out of the cell on her own and must wait for workers to chew away the wax at the tip of the cell. Once this has been removed, the new queen can cut away the remains of the pupal case.

By the time the new queen emerges, her mother will normally have left the hive with a swarm or will have been removed by the beekeeper. The only exception is the process of supersedure where the mother queen and her daughter can coexist for some time. The new queen will now move rapidly though the hive to find and kill any other queens. She will be attracted to other queen cells in the colony by the pheromones given off by the other queens in their queen cells and, with the help of workers, will try to sting any viable queens developing inside queen cells.

If the colony has swarmed and is still strong and numerous it may be able to swarm again. In this case the workers will pre-vent the new queen getting to other queen cells by clustering round them. When the new virgin queen is ready to leave the hive with another swarm the remaining workers will release another virgin queen. On occasion, the workers will release more than one virgin queen but prevent them getting together, so that they do not fight and swarms may leave the hive with more than one queen.

Sometimes the beekeeper or the bees will release all the virgin queens at the same time. The result can be a fight between the queens to leave just one survivor. These fights are ritualised and resemble a wrestling match. Once the stronger (and fitter) queen is able to hold the other down she will move so that she can sting the weaker queen at the back of the head and kill her. The fight is so designed that the surviving queen is rarely damaged and able to become the productive head of the colony.

New virgin queens are quite small and very hungry. Their abdomens tend to be more pointed and shorter than fully

Supersedure

A queen could live up to five years, but normally she will be replaced by the colony after perhaps two to three years as her pheromones start to reduce.

This replacement is a natural process called supersedure and beekeepers can take advantage of this by using any spare queen cells that the colony produces.

A crowded entrance to the hive as bees prepare to swarm.

mature queens because their ovaries have not fully developed and started to produce eggs. The newly emerged virgin will move swiftly about the hive and feed herself to put on weight. During this stage, which lasts for about four days, the workers in the colony seem to take little notice of the virgin queen and it can be very difficult to find her when inspecting the bees.

Mating time

After these four days the new queen will have developed her tergite and mandibular glands and she will start to produce pheromones that are very attractive to the workers. From this point and for the rest of her life the workers will feed the queen. They will control her and care for her as well. She is now sexually mature and ready to mate with the drones.

Mating takes place outside the hive. When the workers consider that she is sufficiently mature, the queen will be forced out of the hive. She will fly to an area where drones collect (known as the drone collection area) and then be chased by a number of drones. She will stay in this area for up to 10 minutes and will mate with a number of drones during that time.

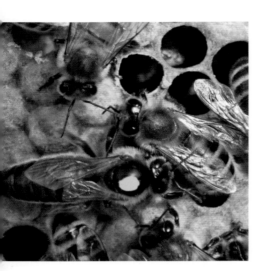

Above: Workers attending a marked and wing-clipped queen. When she has successfully mated she will have three to four years of sperm stored away in her spermatheca.

Right: A mature drone. Look at the size of the drone by comparison to the worker bees in this image.

Should the queen not be able to mate on one of these flights or only mate with a few drones, the workers will encourage her to go on more mating flights until she has mated with many drones. Normally a queen will mate with up to 20 drones before she completes the mating process.

With each mating she will receive many millions of sperm that migrate through her vagina (*bursa copulatrix*) and eventually arrive in a special organ called the spermatheca where the sperm will be kept alive until required to fertilise an egg. The spermatheca can hold up to eight million sperm but during the mating process the queen may receive over ten times this amount from many drones. She mates with so many drones to ensure her workers will have many fathers so the colony will be genetically diverse and will have a greater variety of capabilities.

The queen will hold sperm from all the drones she mates with but her body will reject up to 90 per cent of the sperm from each drone. This seems wasteful but should the queen only be able to mate with one or two drones, she will still have enough sperm collected to last her a lifetime of egg laying.

Protecting the queen

Leaving the hive to mate puts the queen at great risk from being eaten by birds (particularly house martins, swifts and swallows) as she is relatively large and slow-flying and therefore easy pickings. It is now thought that when the queen leaves the hive, she is accompanied by a small entourage of worker bees. They protect her and guide her to where the drones are waiting. The entourage then waits until she has finished the mating process and will guide her back to the hive. Birds are less likely to find the queen in a group of bees and because the worker bees have been out of the hive foraging, they will be more aware of the surrounding geography and the location of drone collection areas.

The virgin queen has about three weeks to mate with the drones. If, for any reason, she is unable to mate during this period her ability to mate is lost and she will only be able to lay unfertilised eggs. If this happens, the colony will produce no more workers, as each unfertilised egg will produce a drone. The colony is then doomed as it will be unable to create another queen and the current laying queen is of no value to the colony.

The queen's egg-laying life

A queen has the ability to continue to lay eggs ad infinitum for she manufactures and grows eggs all her life.

An unmarked dark coloured queen, that will be more difficult to spot for a novice beekeeper, being attended by a group of workers.

The mated queen

Once she has mated with a number of drones the queen settles down in the hive to lay her eggs. Barring any accidents, the workers will not allow her to leave the hive again unless the colony swarms. The lifespan of the queen depends on many factors. Queens do not normally die from old age but at some point the workers determine that she is past her most effective breeding time and will replace her with a new queen. The workers will take a number of different approaches to requeen depending on how numerous they are, how much food is available and the general physical condition of the queen.

Top: Supersedure cells. When the old queen is starting to fail, the workers start the process of raising a new queen, and thus eventually replace the older matriarch.

Centre: A poor frame, showing comb damage, drone and worker brood, distorted cells and queen cups. This frame should have been replaced last season.

Colony replication

If the colony is full of bees and there is plenty of food, the colony will split into two colonies with a good chance they will both survive through the next winter. The old queen will leave with about half the workers in a swarm and find a new home. The rest of the workers will create a new queen from a new worker larva and rebuild the colony. In the wild, colonies normally build to this size and capability over a period of two to three years.

Supersedure

It may happen that the colony has an ageing queen but is not strong or populous enough to swarm. In these cases the colony will supersede the old queen. They will develop one or two queen cells but not reject the old queen. When the queen cells hatch, one of the virgin queens will survive and go on mating flights, but the old queen will stay in residence. The outcome can be that there will be two queens in the colony (a mother and her daughter) who are laying eggs alongside each other. Later in the season (usually in the autumn) the old queen will be rejected by the workers and die, leaving the young one to continue with the colony. Supersedure is a process that the bees can handle on their own with no intervention from the beekeeper.

Beekeepers recognise two forms of supersedure.

1. Perfect supersedure where the mother queen and daughter queen coexist for some time and will both lay eggs. It is called 'perfect' because in this scenario there is no let-up of egg laying in the colony. Usually the mother queen's egg laying rate gradually diminishes whilst the young daughter queen's egg laying rate increases. There will come a time when the daughter queen's laying rate is sufficient for the colony and at this stage the mother queen will be disposed of by the colony leaving just one queen to continue the colony.

2. Imperfect supersedure is where the old queen is dispatched before the new queen starts to lay eggs. This is imperfect as the colony experiences a gap in egg laying and a consequential loss of workers for a period until the new queen reaches her optimal laying rate.

Above: A large number of worker bees clustering at the entrance is often a sign of an imminent swarm.

Left: Workers may cluster like this for a day or so before finally setting sail to find their new home. The observant beekeeper will take prompt action to prevent the mass exodus of bees by ensuring that the colony has sufficient room in the hive.

Drone development

Drone cells are often located along the edge of the brood nest area and quite often in the corners of a frame.

Placing the unfertilised egg in these positions on the frames helps development of the much larger drone pupa.

The drone larva 'sheds' its skin (the old skin is absorbed and the new softer skin can expand before it hardens) during development within the cell (as do workers and queens) and when the process is complete the cells are capped over after about seven days (larva) with the distinctive domed capping where they spend another 14 days (pupa) before emerging.

A colony invests considerable effort in ensuring drones are fed well as larvae and after emergence for another couple of days, then they are left on their own to fend for themselves.

Swarming trigger

In managed colonies the beekeeper is able to support the workers when it is clear that the queen is getting too old to remain with the colony. There have been a number of theories developed over the last 150 years to identify the triggers that mean a colony is about to swarm – see page 123.

The swarm with the old queen will establish a new nest site. Then, at some point, usually before winter, the old queen will be replaced by her daughter queen. This supersedure ensures that the swarmed colony has a new and active queen to help build their numbers in the following season.

The outcome of the process of queen replacement is that a colony of bees can live and survive for many years but, the queen may only live for up to five years whilst workers will live for a few weeks. It is for these reasons that a beekeeper must consider the health and wellbeing of the colony above the individual needs of a single worker or queen.

Drones and mating

The drone bee will emerge from the capped cell 24 days after the egg was laid. At this stage his testes will have already produced and matured all the sperm that he will require. However, he is not yet sexually mature or sufficiently strong to consider mating. He will stay in the hive and consume pollen and honey for about 10 days. During this time he will develop so that he is able to produce sexually attractive pheromones and will go on a number of short flights to strengthen his muscles and orientate in the locality. During this time he will also learn how to beg food from workers and will be rewarded by them.

Once he is mature he will go on mating flights and assemble with other drones in a collection area to wait for the arrival of a virgin queen. In the collection area the drones fly around in a fairly haphazard way for about an hour until a virgin queen arrives. As soon as the drones sense her by her pheromones they chase her and the strongest and fastest-flying drone will catch her first and mate with her. The bad news for the successful drone is that if he does catch the virgin he will mate with her but the process paralyses him and his endophallus (penis) remains in the queen as he drops to the ground and dies. It appears that a drone will either mate once with a queen and then die or live for a couple of months in the hope of mating and then either die from exertion or be forcibly removed from the colony by the workers as winter approaches. In either instance drones do not live and survive through winter.

Drone duties

Whilst drones are waiting for the opportunity to mate with a virgin queen they appear to perform very little work in the hive, but having drones in the colony seems to help maintain its cohesion. Drones seem to be able to enter any colony and will drift from one home to another with little opposition whereas workers normally remain with the colony that nurtured them as larvae. There may be genetic value in drones moving from colony to colony as this broadens the gene pool in a particular area and helps to disperse any positive survival advantages that the drones possess.

Drone collection areas

Drone collection areas (DCA) are well-defined areas in the air where drones assemble to wait for the arrival of the queen. Much research has been conducted into these areas to understand the characteristics that create a drone collection area and how they are distributed.

They all have a number of significant features:

- They appear to be located where there is a slight updraft presumably to reduce the effort of flying.

- They are normally about 10 metres above the ground.

- They appear to be located in the same place from year to year.

- They contain between 1000 and 10,000 drones; fewer drones and the group will disperse, a greater number and the drone collection area will break into two or more collections. Drones produce pheromones that attract other drones. When they collect in a drone

A flying gamete

The drone comes from an unfertilised egg and possesses only half the genes a normal adult bee would have.

He has no father and has only characteristics passed on by the queen, hence the term gamete.

Drones from the same queen are therefore very closely related and if they did not move from colony to colony it is likely to aid the genetic diversity in any drone collection area.

Without these two factors, recessive and undesirable characteristics could develop in populations with potentially disastrous consequences.

Above: The bees have built drone comb on the bottom of this frame – see the difference in cell size. Beekeepers may sacrifice this drone comb in order to reduce the Varroa load in their colony.

collection area the pheromone scent will be strong. This feature may be used by the queen and her attendant workers in order to locate a drone collection area.

■ Not all known drone collection areas are used every year and this probably depends on the number of drones living in a particular area in any one year.

Drone collection areas are difficult to find although some people have said that they have heard a buzzing noise and looked up to see a large number of 'bees' flying about. It is possible to find a drone collection area by tying a lure, covered with the pheromone produced by virgin queens, to the end of a long fishing rod and walking about in an area with it. If the lure enters a drone collection area a 'comet'-shaped collection of drones will chase it and try to catch the lure. Using this technique, drone collection areas have been found and mapped. This has shown that the outline of the drone collection area is sharply defined. If the lure leaves the drone collection area, the drones will cease to chase it. However, if the lure is lowered once the drones are chasing it, the drones will follow it to the ground.

Obviously drones cannot remain in the area all day: they move to these areas at about midday and stay there for about an hour. It seems that virgin queens will fly to the areas once the drones are present, but the queen is only able to stay out for only a few minutes. It may be that drones go to the area and stay until their energy reserves are nearly exhausted. They then return to the hive and feed before going out again.

In addition, the weather needs to be suitable in order for drones to gather and virgin queens to venture out. Ideal conditions are warm (greater than 20°C), calm and preferably sunny. If the weather is inclement for about three weeks when the virgin queen is ready to mate she may not be able to do so, resulting in a queen unable to produce workers (from fertilised eggs).

It is also essential that there is a virgin queen that needs to mate. Often when a colony is without a queen and is generating a new one, many drones will collect in the colony as if waiting for the virgin queen to mature. Once the virgin queen has fully mated

At the top of the valley the breeze blows softly and drones are sometimes seen awaiting a queen on a warm summer day.

and is laying eggs, the number of drones in the colony will dramatically reduce as they depart to a new colony.

The mating process

Virgin queens and drones mate on the wing and the process is designed so that the queen can mate with a number of different drones during a single mating flight. Because of this, the strongest drones have an advantage over other drones. As the queen flies through the drone collection area the drones will sense her pheromones and immediately be attracted to her. She will leave a pheromone trail and the drones are able to detect its direction as it becomes stronger and gets nearer to the queen. They chase after her, in a characteristic comet shape trail.

As the drones approach, the queen opens her copulatrix (the entrance to her vagina) and the first drone to reach her will hold her abdomen with his fore legs and arch his back so that the end of his abdomen is facing the copulatrix. He then contracts his abdomen in an almost explosive manner (some people claim to have heard a crack when this occurs). The contraction everts his endophallus (penis) that is normally not seen and held inside his abdomen. The motion forces semen inside the seminal vesicles in his abdomen into the queen's vagina and a plug of mucus follows to stop the semen draining away. This massive action also breaks off his endophallus and he is immediately paralysed and falls to the ground without his endophallus and semen. The successful drone dies having performed his important role.

This is not the end for the queen, as she needs to mate with several drones. She continues to fly and the next most powerful drone will catch her and repeat the process. The next drone has to remove these with his hind legs before copulating with the queen. This process may be repeated up to ten or a dozen times after each mating encounter in one flight or the process may be aborted if the queen can find no additional drones.

Should the drone chasing the queen be closely related to her, that is, he has the same mother queen, the drone is able to sense this. He will be repelled by the queen and give up the chase. He will then wait for another queen to come along. Even though

Multiple matings

Recent research suggests that multiple mating produces a better quality colony because of the greater genetic diversity in the subsequent workers.

Tell-tail signs of mating are white strands left behind by a drones endophallus after copulation. These can just be seen in this image.

virgin queens and drones may come from the same colony (although the drone does not stay in the colony where his parent queen resides), they show absolutely no interest in each other either in the hive or outside the drone collection area.

Rearing queens

Beekeepers have two options when deciding how to deal with the queen. They can either allow the new queens to take control of a colony after the old queen has departed or practice an element of selection so that certain qualities of the colony can be preserved or improved. The very nature of natural mating means that the selective breeding that is employed with other animals is not possible without a completely artificial approach to mating. Instrumental insemination, as it is known, is normally only done at research institutes or by dedicated queen breeders who supply particular types of bees to other beekeepers. This approach, though interesting, is not practised by amateur beekeepers and is not covered in this book.

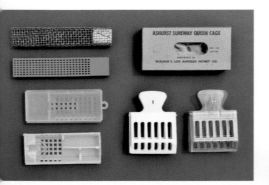

Top: A selection of queen cages, from the early cardboard traveling cage, to modern plastic cages and two 'Butler' style cages top left.

Right: A queen rearing workshop held in 2009 during the annual Spring Convention at the National Beekeeping Centre, Stoneleigh Park in Warwickshire.

When starting out with one or two hives, most beekeepers will allow their colonies to raise new queens when they are ready. Using a simple method of swarm control they will allow the bees to go through the natural process of splitting into two colonies where one will have the old queen and another will raise a new queen. Once the process is complete and the new queen is laying successfully the two colonies can be reunited with the new queen. In this way the colony can continue with a new queen without the loss of bees and possibly worrying your neighbours. Swarming and swarm control is covered in Chapter 7 which shows simple methods of replacing an old and failing queen with a new one.

Desirable colony characteristics

A beekeeper with a number of colonies will be able to assess the individual characteristics of each colony and determine which suits him best. These are the colonies that beekeepers with the knowledge and capability will choose to raise new queens. The qualities a beekeeper may choose can vary but most would suggest that the following are critical for any beekeeper:

- Health or disease resistance and
- Gentleness.

It is no pleasure to keep bees that are so difficult to work with and so defensive that they will try to sting any object that gets close to the hive or chase away anyone near the hive. Similarly it is not good for the bees or the beekeeper to propagate colonies that are susceptible to any disease. The colony cannot build up as it should and will have great difficulty surviving its first winter.

Other characteristics that should be borne in mind are:

- A low propensity to swarm may be useful in urban areas where swarms can be a nuisance to the public.

- The production of large quantities of honey may be a characteristic that commercial beekeepers wish to select for but for most amateur beekeepers this is a secondary consideration.

Learn how to rear queens

A number of queen rearing courses are held throughout the country by local associations during the active beekeeping season and details will be found on the BBKA's website.

Queen rearing is a fascinating and often frustrating part of bee improvement and when you have had a season or two of beekeeping you will probably want to attend a course to help improve your colonies for disease resistance, temper, yield or other characteristics.

Preparing to cut out a suitable queen cell for 'planting' into a colony of feisty bees.

- A colony that survives winter well and builds a strong colony in spring.

- Bees that are able to defend the colony from predators, such as wasps.

In order to raise queens that will pass on desirable characteristics, it is important to keep records of the performance of your colonies. With only one or two colonies, it may be difficult to recognise a colony that has the most suitable characteristics. If this is the case you should work with local colleagues so that the selection process is able operate over a larger number of colonies and the general characteristics of colonies in the area can be improved.

Recording details of each visit and the state of the colony is vital for successful queen rearing.

Breeding virgins

The selected colony with the desired characteristics should be encouraged to raise new queens.

Above and centre: Three supersedure queen cells, the larger of which may be carefully cut out and inserted into another colony, as in the centre image. The other two queen cells are of poor quality and will be destroyed.

There are many ways to do this:

1. One of the simplest methods is to **remove the queen from the colony**. The queen is removed along with two frames of bees into another hive, known as a nucleus hive (this will be explained in more detail in Chapter 7). The now queen-less colony will start to

produce new queen cells and feed the young larvae in order to produce a number of queens. Ten days after removing the original queen there will be a number of sealed queen cells in the colony and these can be cut out of the frame and placed in small queenless colonies that will care for the queen cell and the resultant queen. The virgin queens will mature in these small colonies and, when ready, will fly to mate with the local drones. When this process is complete, the virgin queens will start to lay eggs that will result in the creation of their own workers.

2. **The Miller method:** With this approach you need to find a frame of eggs and very young larvae in the desired colony. Cut the lower edge of the wax comb through a section with very young larvae to create a number of 'v's. The cut edge should pass through a line of cells with these larvae and leave them exposed.

   Pinch out every other cell along the line and then place the frame into a colony without a queen and with many young bees. The young bees will preferentially select these exposed larvae to make new queen cells and after a few days there will be a line of queen cells along the cut edge. These can be removed and, as before, placed into small colonies to mate and develop into productive queens.

3. **Punched cells:** with this method a frame of very young larvae is removed from the selected colony. A cell is selected with the young larva in it and from the other side of the frame, the cell with the egg is removed using a cutting tube a bit larger than the cell. This wax cell is then mounted onto a wooden former that is held horizontally in the hive on a dummy frame. The former can take up to 10 'punched' cells. The former is placed into a colony of young bees that do not have a queen. The bees will rapidly adopt the punched cells and build them into queen cells.

The Miller method. A frame of wax with young larvae has been cut in two 'v's and the bees have started to make queen cells along the wax edges.

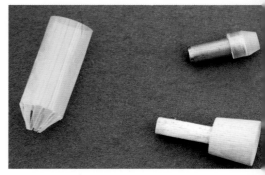

The kit required to punch cells of young larvae before mounting them on a bar for growing on to queens.

4. **Grafting:** This is a technique use by many more experienced beekeepers. Once again a frame of very young larvae are removed from the donor colony. The cells are looked at to find larvae suitable to raise as queens. Using either a fine paintbrush or a special grafting tool the larva is lifted out of the cell and placed gently into a plastic or wax formed cell that is designed to be mounted onto a horizontal bar in a queenless colony with many young bees. If the grafting process has been done without damaging the larva then the workers will build a queen cell and raise a new queen. The fully formed queen cell can then be placed into a small mating colony to get the queen mated and ready for moving into a productive colony.

Top: Using a 'Chinese' grafting tool to lift out a very young larva.

Below: A bar of young queen cells that have 'taken' one day after grafting 1-2 day old larvae.

Centre: A grafting frame with queen cells lifted out of the brood box to check on development. The bees will need to be brushed off to make a check on those cells that have 'taken'.

All these methods are difficult to describe sufficiently well on paper. The best way to learn is to watch an experienced beekeeper demonstrate these methods.

The colonies should be allowed to increase in size until it is possible to assess their qualities. This can take about two months. Having reached this stage the queens with the desired characteristics can be introduced into the colonies that have old queens with either unwanted characteristics or which are soon going to be replaced by the queen-rearing efforts of their own workers.

In effect this will produce a number of virgin queens from a single colony. These queens will inherit similar characteristics to the workers in the colony. However, without any control over the drones that subsequently mate with the queen there is still some risk that adverse characteristics may be seen in the colony headed by the new queen.

Rearing drones

Controlling the drones that mate with your queen is complex and only attempted by very experienced beekeepers. The alternative is to instrumentally inseminate queens with semen extracted from specific drones. Once again, this is complex and normally only done for the purposes of research or by professional beekeepers who sell large numbers of specialised queens. It is not for the amateur beekeeper who is just starting out.

Marking queens

In beekeeping circles marking queens is a contentious issue. A queen is marked with a coloured spot on the back of her thorax using a special marking pen or quick drying paint. To do this, she must either be trapped in a cage, so that she is unable to move, or be picked up and held by the thorax between the thumb and forefinger. These actions are considered by some to stress the queen and to mutilate her. The reason it is done is to make finding the queen much easier when the colony is inspected.

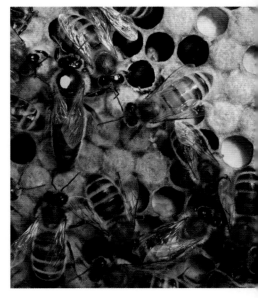

A well marked queen is easier to spot in a crowded colony.

Although it is not always necessary to find the queen, there are certain manipulations that are simpler and less stressful to the colony if she can be found. Some beekeepers use the marking to find the queen quickly and to place her in a queen cage whilst the rest of the inspection continues to ensure that she is not damaged. Natural and 'organic' beekeepers are set against this practice, as they believe that it involves too much interference by the beekeeper. On the other hand some very experienced beekeepers do not mark their queens purely because they are so good at their craft and have developed techniques to manage

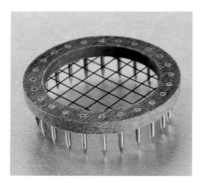

A queen marking cage known by many beekeepers as a 'crown of thorns'.

Top: Holding the queen gently at the Thorax or legs – never by her abdomen.

Bottom: Carefully clip off the end of one wing only by about a third of the wing length.

colonies without sight of the queen. If on a rare occasion, they do need to find her they are able to do so easily. For the average amateur beekeeper it is quite difficult to find a queen amid 50,000 workers who seem to be able to hide her whilst going about their normal behaviour on the comb.

Every beekeeper must decide whether or not to mark the queens, but it will make the early years as a beekeeper easier, and there is always a delight in seeing the queen in the colony even if it is not necessary.

A special cage is used to immobilise the queen. Once the queen has been located, the cage is gently placed over her and then pushed into the wax comb until the queen is just unable to walk. This must be done gently and with care so that the queen, and especially her abdomen, is not crushed. Once she has been immobilised it is an easy job to dab the marking pen on the back of her thorax and lift the cage slightly so that she can move but not leave the cage. Once the paint is dry the queen can be fully released and watched to ensure that she is unharmed and accepted by the colony.

Normally there is no problem with this, but should the workers start to attack the queen because she no longer smells like their queen, she needs to be caged again in a queen introduction cage so that she can be introduced more slowly (details of how to introduce a queen to a colony are given in Chapter 7). Once the beekeeper has become more adept and confident in handling bees it is possible to pick the queen off the comb using her wings and then move her to the other hand so that she can be held by at least two legs on one side of her body between thumb and forefinger. This effectively immobilises her and she can then be marked without the use of cage. After marking she is lifted off the beekeeper's fingers by her wings and placed gently back on the comb.

Clipping queens

Clipping a queen is probably even more contentious. This is the process of removing about a third of one forewing of the queen with a pair of scissors. The reason for doing this is that it prevents the queen from flying away. If she attempts to fly, the clipped wing unbalances her so greatly that she will spiral to the ground. If both wings are clipped the queen will still be able to fly but it will take far more effort. Once a queen has been mated the only time she might need to fly is if she leaves the colony with a swarm. A clipped queen will not be able to leave with the swarm and will return to the hive. A clipped queen will give the beekeeper more time to check the colony and ensure that it does not swarm of its own accord.

During the swarming season (April to June) it is necessary to check the colony once a week to find evidence of preparations to swarm and then to take the necessary action to prevent possible nuisance to neighbours and the potential loss of bees from the colony. Once the queen's wing has been clipped, the weekly inspection can be cut down to once a fortnight. This reduces the intervention by the beekeeper and reduces the stress on the colony. Furthermore, while the normal swarming season is between April and June, swarms have been known to occur in every month of the year.

When managing your own colonies you have to weigh up the benefits and disadvantages of clipping. The queen is definitely mutilated with part of one wing removed but she shows no sensation of pain even though there are sensory organs on the wings. Queens that have been clipped go on for another two to three years performing their natural function. There is some concern that clipping a queen's wing may interfere with some communication through wing vibration between the queen and workers. This mutilation must be balanced against the reduction in frequency of inspecting a colony from once a week to once a fortnight. This cuts down the overall stress on the whole colony as well as the risk of a swarm leaving the hive and causing a nuisance to the public.

Knocking bees off the frame by banging one hand against the other holding the frame.

Swarms

Swarming is a natural process for honey bees and its purpose is to reproduce colonies of bees. If bees did not swarm the number of colonies would diminish over time. Some would die, either from disease or starvation, until no more colonies were left.

A large swarm has collected on the lower branches of an ornamental tree. They are not always as obliging as this and often settle in difficult places. This swarm was easily collected by a neighbouring bee-keeper.

When a swarm occurs, a colony will split into two viable groups. One will stay in the original site while the other flies off to find a new nest site. When managing honey bees the beekeeper must acknowledge that swarming is a necessary activity and that bees will always attempt to swarm when the conditions are right.

The beekeeper's job is to help the bees through the swarming instinct while ensuring that those that do swarm do not become a nuisance and a threat to neighbours or end up making a nest in an inappropriate location such as a chimney or cavity wall.

The swarming process

During the spring and early summer a colony of bees will expand rapidly in numbers and, if the conditions and food supply are good, will reach a stage where the colony becomes overpopulated. Scout bees will go from the hive looking for new nest sites in the vicinity so that when swarming occurs some of the bees will have identified possible locations for the new nest.

Workers will select a number of brood cells – usually situated towards the periphery of the wax combs – and will either modify these so that the larva can hang downwards in the cell or they will make a 'queen cup' which resembles the shell that holds an acorn. The queen will then be encouraged to lay an egg in the modified cells or cups. There can be as many as 10 or 20 such cells when the colony is in a condition to swarm. At this point the queen will receive less food from the workers and her laying rate will slow down. The queen's weight will reduce and her ability to fly will be improved when the time comes to swarm and leave the hive.

When the eggs hatch in the modified cells and cups the workers will feed the larvae with royal jelly to produce queens rather than workers. The cells will lengthen in a downwards direction as the larvae get bigger. After five days workers will seal the ends of the cells with wax and the larvae will start to pupate, eventually emerging as adult virgin queens. Once the first queen cell is sealed, the colony will become very agitated and some of the bees will rush through the colony to disturb it to initiate the swarming.

Initial swarm site

Soon after, many bees (over half the colony) will fly out of the hive and force the old queen to leave with them. There will be a great disturbance near the hive with many thousands of bees flying about in a seemingly haphazard way. Somewhere in the melée will be the queen, guided by the other bees. The workers produce a pheromone from their Nasonov gland which helps to keep the swarm in a close formation. The swarm will eventually find a resting point nearby, usually a branch on a tree about two to three metres above ground.

Swarm time

In the south of England in an early spring swarms may occur in early March whereas in the north swarms may typically be later in May or June.

There's an old rhyme that goes:

> *'A swarm of bees in May is worth a bale of hay'.*
> *'A swarm of bees in June is worth a silver spoon'.*
> *'A swarm of bees in July is not worth a fly'.*

Swarms are a useful source of new bees for the beekeeper who should ensure they are fit and healthy before introducing them to his apiary

The beekeeper has destroyed the queen cup by opening it out with a hive tool.

Controlling swarms

Swarms can be the cause of considerable distress and a nuisance to neighbours, although in themselves they are relatively harmless.

The noise and sight of many thousands of flying bees can seriously upset the best of friend-ships, particularly if they settle in a chimney or outhouse. The swarm can look and sound very menacing to the uninitiated.

Beekeepers have a duty to take care of their bees and should take the trouble to ensure their precious stocks do not escape and cause a nuisance.

The swarm will start to collect here in a loose assembly of bees, each holding onto other bees to form a cluster. The queen will be with the cluster although she is able to move freely through the mass of other bees. Once the swarm starts to settle, bees can be seen flying between the swarm and the old nest for a few minutes so it appears that they are balancing the two colonies. In the first stages of swarming the old colony may be depleted of bees and, during the balancing process, bees will return so that the old colony still has enough bees to raise the new queens and take care of the hive.

Selecting a new site

After a few minutes all goes quiet and the colony in the hive continues its normal duties while the swarm settles down on the branch. Scout bees will be seen leaving and returning to the swarm. The scouts are now inspecting all these sites and deter-mining which would be most suitable as the new home. Having inspected a site each scout bee will return to the swarm and perform a dance on the swarm to indicate when the home is. If the site is good the scout will continue the dance and attract other scouts to visit the home. Eventually, by a process of elimi-nation and persistence, all the scout bees will all indicate that one site is better than all the others. At this stage they will all perform the same dance and trigger the swarm to leave the branch and travel to its new home.

The process of checking out a future home for the colony can take any time from about an hour to a day and sometimes the swarm will not find a new home. In this last case the colony stays where it is and builds a nest in open air. It is invariably doomed and will die at some point before or during winter.

Final flight

During the flight to the new home the scouts will fly through the swarm leaving a pheromone trail to guide the other bees and, when the colony reaches the new home, there will be other scout bees at the entrance to guide the swarm in.

If at any time in the whole process the swarm perceives that the queen is not present, the swarming process collapses and it will look for the queen. If she is not found, the swarm will return to the old nest and the impetus to swarm will be gone.

The old colony

Once the swarm has left, the remaining bees will continue their work foraging and caring for the queen cells in the nest. Over the next eight days the queen cells will mature and, about a day before they are ready to hatch, the workers will chew away the wax covering the pupal cases at the bottom of the cells so that

Inspect to avoid a swarm

Regular inspections during the swarming season will help prevent an unexpected swarm.

Once the colony starts to become overcrowded and the weather is warm and dry, bees often begin to prepare for swarming!

A swarm has conveniently settled on a fence and the beekeeper is using a skep to collect the bees for rehiving

the queens may cut their way out. The first queen to come out will walk rapidly round the nest and feed on the pollen and nectar. She will be looking for other queens in the same nest. If she finds one, a fight will ensue, and one queen will die. She will then find queen cells that have yet to be opened and, with the help of the workers who will remove some of the wax covering the pupal case, sting the un-hatched queen in the cell.

Virgin selection

Old queen cells can be found with a hole in the side where this process of sororicide (sister killing) has taken place. The only defence that the victim queen has in this situation is her ability to make a high-pitched sound, known as 'piping'. When a queen pipes it appears that the noise momentarily stops the workers from tearing down the cell. A queen also uses piping as a way to alert other queens to her presence. It seems that the victor and the defeated queen in these battles are keen to get the fight over and will actually attract each other to a fight. There is no possibility of hiding and hoping that the other queen will go away.

Colony size

The old colony now has a young virgin queen and sealed worker brood. No eggs are being laid and there has been no possibility of any further egg laying for the past week. The older bees are now dying and, until the new queen is fully mated, no more new eggs will be laid. This can take a further three weeks. In all, the old colony will have no eggs laid and new bees produced for nearly six weeks.

Prior to preparations for swarming the queen could have been laying up to 1500 eggs a day so there is a significant reduction in the size of the colony.

Once the new queen starts to lay eggs, her pace increases rapidly and she will hopefully ensure that the colony is able to survive. The old colony is at greatest risk during the period between the old queen leaving and the new queen starting to lay eggs. If, for any reason, the queen does not return from a mating flight or she is not able to mate successfully, the colony is probably doomed.

Top: Queen cells close to emergence. See how the workers have chewed the tip of the cells smooth in anticipation of the queen emerging.

Bottom: A young queen looking for a suitable cell in which to lay her egg.

The beekeeper can help the colony if this does happen, but without assistance, the colony will be die out. By introducing a frame of eggs from another colony (one you own or supplied by a beekeeping colleague) the workers have a second opportunity to raise a new queen and survive as a viable colony through the coming months.

Casts from the old colony

In times of plenty, colonies build up strongly in the spring. Some colonies get carried away by the urge to swarm. After the original (prime) swarm has left, the workers prepare one of the queen cells so that the virgin queen can emerge. If all of the queens are allowed to hatch at the same time they will fight until only one is left. The 'last one standing' will then start to go on mating flights and take over the duties within the colony.

Should the workers only allow one queen to emerge, this queen will swiftly sense that there are other queens in queen cells and try to sting them through the cell walls. She does this with the cooperation of the workers who will remove some of the wax covering the queen cell. Alternatively, the workers are able prevent this and to protect the other cells by covering them with their bodies and preventing the queen from getting near the victim's queen cell.

Eventually, once the 'released' queen has matured fully, she will swarm with around half of the remaining bees. This is called a cast. The reduced number of bees in the cast lowers the possibility of it surviving through the first winter. Again, the remaining workers in the old colony will release another queen and the process will continue until there are only a few hundred bees are left in the colony. Even in times of plenty, any subsequent casts after the first cast are unlikely to survive for any time and it seems strange that the colony should effectively sacrifice itself in this way.

Colony replication

The process of swarming will usually produce two colonies from one and allow the number of colonies to increase. In normal circumstances this will ensure that any colony losses from

The first cast from a colony after the prime swarm has left. It can happen that more than one cast issues from the remaining bees in the colony and it is not unknown for there to be four or five casts in a poorly managed hive.

previous years can be recovered. Should the conditions improve such as better food supplies and warm summers, the total number of colonies will be able to increase. If conditions worsen overall then the total number of colonies will decline. A swarm may not be able to survive if adverse weather prevents the swarm building up colony numbers or if an outbreak of disease strikes in the area.

New colony

Once the colony has reached its new location, the workers must start to build new wax comb so that the foraging force can collect honey and pollen to keep the colony alive. A large quantity of wax comb is required because until full-sized cells are in place the queen cannot start to lay eggs to re-establish the colony. Throughout this period, the tireless workers will be dying from exhaustion. Should the colony suffer from an extended period of bad weather or an outbreak of disease the old workers will die off before the new workers have been produced and the colony die out.

Supersedure

Once the swarmed colony establishes a new nest with plenty of bees, the queen that came from the original hive is often superseded. A swarming queen can often be two or three years old and coming to the end of her useful life. Rather than simply killing the old queen the colony will raise a new queen from one or two queen cells near the centre of the brood nest.

Top: An emergerged queen cell.

Bottom: Two supersedure cells that unless the beekeeper needs them to requeen another colony may need to be destroyed.

Centre: Supersedure cells that are almost hidden underneath bees in the centre of the image and therefore easy for the novice to miss in a very crowded colony.

Once the virgin queen emerges from the queen cell she will be cared for by the workers and encouraged to go on her mating flights. She will eventually settle down to lay eggs and will often do this alongside her mother queen. As the season moves towards winter the older queen is cared for less and eventually dies, leaving a new, young and well-nourished queen to take over all the egg-laying duties.

Triggers for swarming

Over the last 150 years, there has been a great deal of research to determine just what triggers bees to swarm. In the 1890s a German scientist called Ferdinand Gerstung deduced that as a colony built up in size, there came a point when there were just too many young bees producing brood food. At this stage the colony would swarm, as there were plenty of bees and food available to see the two colonies through this difficult time. In the 1920s G. S. Demuth, an American researcher, identified the trigger as overcrowding in the nest. He believed that the bees had run out of space and needed to swarm to reduce the congestion.

However in 1953, a Briton, Colin Butler identified that queens produced a pheromone known as 'queen substance' and if the quantity of this was reduced the workers would initiate the swarming process.

Queen substance is now identified as one of the most important pheromones produced by queens. It is passed about the colony when the workers touch the queen and then each other. Should the substance not be available, the bees will rapidly start swarm preparations.

Butler found that the amount of queen substance a queen produces drops by half for each year of her life. If the colony was crowded the ability of workers to distribute the substance through the colony was diminished. Butler's work was able to explain the experimental evidence seen by both Gerstung and Demuth and the reduced quantity of queen substance is now considered to be the prime trigger for swarm preparations to begin.

Swarming in progress. Workers are waiting for news of a suitable new home before setting off.

Beekeepers can simulate the effect of no queen substance by removing the queen from a colony. Within a few minutes the workers will recognise the queen is missing and search for her. Soon after the workers will try to build a number of queen cells from existing cells with very young larvae. It is suggested that some of the chemicals in queen substance are quite volatile and, after about 30 minutes, half of the pheromone elements present in the colony would have evaporated. Other components in queen substance have been shown to have a longer term effect on the workers and ensure that their ovaries are incapable of producing many eggs.

Top: A typical large swarm hanging in a tree. This one should be fairly easy to collect.

Right: Adding supers above the brood box is a simple way to aid swarm prevention. Here a large colony has six supers added – no wonder the beekeeper is smiling.

Swarm prevention

This is the process by which a beekeeper seeks to remove the urge to swarm. As described above, should the colony become congested bees will swarm. If the beekeeper tries to keep an old queen in a large colony of bees the likelihood of swarming is also increased. As April passes into May it helps to reduce the swarming impulse if the bees are given plenty of space in the hive.

Extra space can be provided by adding more supers. This allows the bees to move into a new area and relieve congestion in the brood chamber. If you are inspecting a colony on a sunny day it may seem that there is plenty of space in the hive. However, when the foraging bees return to the hive in the evening, the number of bees present can increase by as much as a third.

The second key action is to maintain colonies with young queens. It is generally advisable to replace a queen when she is in her third year in a hive. This is similar to the period that a queen would be allowed to head a colony in the wild. New, fertile queens can be bought during the summer or it is possible to encourage your own hives to produce new queens by using an artificial swarm method (see pages 126-127).

These methods are designed to convince the colony that it has swarmed already and encourage it to produce a new queen in the same way as it would if a swarm had taken place.

Swarm control

Once a colony has started preparations for swarming it is very difficult to prevent the process and every activity to stop the desire to swarm will demoralise the bees rather than allowing them to continue their work. The only action now is to move to a method of swarm control and work with the bees to satisfy their need without losing any bees. In effect, the bees need to believe they have swarmed even though the colony has not left the hive. There are many techniques for swarm control which are usually named after the discoverer who originally documented the process. All methods work on the principle that there are three components in a colony.

Safe swarm collection

The thrill of collecting a swarm of several thousands of free bees and receiving the gratitude of those who believe they have been saved by your skills gives you a warm glow of satisfaction.

But it's easy to forget in the heat of the moment that you should always ensure that you remain safe when collecting a swarm, for bees can land-up in the most difficult places and sometimes at great heights. If you are not sure of your ability to safely collect them, then leave well alone.

Beginner beekeepers checking out the state of a colony.

These are:

- The flying bees (1)
- The queen (2)
- The brood (3)

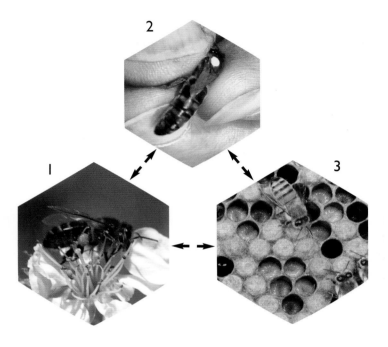

A foraging bee, here collecting pollen from a Californian Lilac (*Ceanothus*).

If one of these three elements is separated from the others, the colony will consider that it has swarmed and the urge to swarm will have been satisfied. In separating one of the elements the beekeeper must be careful not to leave any part of the colony unable to survive. If the queen is removed, she must be left with attendant workers to care for her. Similarly the brood will die without the care of nurse bees so they must be kept together.

Method 1
Remove the flying bees
In this case the queen and the brood are left together and removed to one side. A new queen with some brood is placed on the original hive stand so that the flying bees join the nest. This process is often used by commercial beekeepers as a way of re-queening a colony to improve its characteristics whilst still keeping the foraging force to ensure a good honey crop.

Method 2
Remove the queen

Some beekeepers will remove the queen with a small entourage (about 5000) of workers into a nucleus hive leaving most of the bees and the brood to develop a new queen and continue the colony. The natural break in egg laying and the absence of a queen encourages the workers to continue with the development of the queen cells they have started to create and, at the same time, the foraging bees continue to work and bring in nectar. The outcome will be one large colony with a new queen along with a small nucleus colony with the old queen which can be used as an insurance policy should the new queen not survive her mating flights.

Method 3
Remove the brood

Removing the brood is very similar to natural swarming and is the basis of the most common methods of swarm control as described below. This simulates the situation when the queen leaves the nest with many of the bees leaving the brood with nurse bees to create a new queen.

As a new beekeeper, it is a good idea to use a method that closely mimics the bees' natural tendency to swarm. It is easier to learn just one method of swarm control and then ensure that any additional equipment is available for the time when the bees show signs of swarming. This will prevent last minute panics and will avoid delaying any action needed simply because you do not have enough equipment available.

Above: An unmarked queen.

Below: Larvae and sealed brood with stores of ripening honey

Left: It is probably time for some swarm prevention as the colony is filling and may soon wish to find a new home.

An artificial swarm – the Pagden method

One of the most popular methods of swarm control is often called the Pagden method.

The additional equipment you will need for this method is:
- Hive stand
- Floor
- Brood box
- Crown board
- Roof
- Set of brood frames, either with wax foundation or already drawn with wax cells.

The first action is to move the brood and young nurse bees to one side of the original location and place a new stand, floor and brood box in its place.

- The colony should then be inspected and the frame containing the queen placed in the middle of the new box. Ensure this frame has no queen cells on it.

- Fill the rest of the new box with empty frames.

- Place the queen excluder from the original hive on top together with any supers that were on the hive.

- This new colony is now complete.

- Inspect the old colony (set to one side of the original position) and if there is a large number of queen cells, reduce these to two 'open' queen cells by cutting out any more 'open' queen cells.

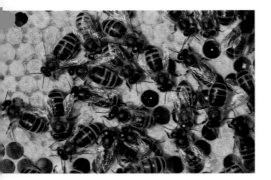

Inspect each frame carefully for signs of queen cells, which can be small at first and concealed by the bees and not as easy to find as the larger queen cell in the lower image!

Many beekeepers suggest that only one queen cell should be left because there is a small chance that with two queen cells the colony developing the new queen might cast and lose some bees. In the author's experience this very rarely happens and, with only one queen cell, there is a danger that this might be damaged or diseased. We believe that the risk of the colony having no queen is greater than the risk of casting and prefer to keep two queen cells for the colony.

The Pagden method of swarm control

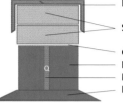

The original hive with queen and bees getting ready to swarm.

Roof
Supers
Queen excluder
Brood box
Frame with queen
Hive stand

Up-turned roof with two supers.

Original hive with queen and bees moved to one side.

New empty brood box with stand placed in the original location.

Up-turned roof with two supers.

Original hive with old frames made up with fresh drawn foundation or empty frames.

New hive with queen on frame and two frames of brood made up with fresh drawn foundation or empty frames.

Original queenless hive with roof added

The new colony is now complete with the original supers and roof added to requeened hive.

The Heddon variation
(After 7 days)

The finished requeened hive now complete and building up after swarm control action.

Move the queenless colony to the other side of requeened hive.

Pagden's method

J.W. Pagden was a British bee-keeper who described this method in 1870.

His aim was to capture a swarm and hive it where the original hive stood and move the swarming hive to one side.

The theory was that it would strengthen the swarm to become a production colony and weaken the old colony so that it would not swarm again.

Early brood, showing sealed cells of brood and stores, together with pollen.

Open queen cells are those that have not been sealed and where a developing larva can be seen. There should not be any sealed queen cells but if they are found they should also be removed.

All the flying bees will join the queen in the new hive (on the original site) while the old hive will contain nurse bees that stay with the brood. Both colonies will be somewhat depleted of bees. Neither colony should have an urge to swarm and the one with the old queen will continue to do most of the foraging. The small colony to one side of the original will eventually raise a new queen and, once mated, she will rapidly make up for the lost bees and build her own new colony.

Above: Two hives marked for a swarm control demonstration.

Right: Inspecting a brood frame for queen cells. Inspections at swarming time should be carried out with care as queen cups can be easily missed. Supersedure cells can be cut out to replace old queens or start another colony and therefore make useful additions to the beekeeper's apiary

Once the swarming period is over (by July in most years) you can decide whether to combine the two colonies to produce one large colony with just the new queen. Alternatively the two can be allowed to develop independently through the rest of the summer and then prepared for winter – a method that allows you to increase the number of stocks held. If there is any mishap, such as the new queen being killed during a mating flight, the two colonies can be combined with the old queen at the head.

In the latter case it often happens that the colony will quietly create a new queen cell and raise a new queen to supersede the old queen later in the season. This is quite natural and results in a new queen without the process of swarming.

Variation on a theme

There are many variations on this approach to swarm control. The most common is the Heddon method, named after another British beekeeper. In this method, after a week has elapsed, the old hive is moved to the other side of the new hive. Flying bees returning to the old hive find that it is missing, and beg their way into the new hive, further reinforcing the number of flying bees in this colony. In the meantime the old colony is further depleted of bees and becomes less likely to be able to swarm once the virgin queens emerge.

There are other methods of conducting an artificial swarm that result in the same outcome, namely a strong colony of foraging bees with the old queen and a weaker colony with a new queen. Neither are likely to try to swarm during the rest of the season.

Heddon variation

Heddon variation to the Pagden method

After a week the colony without the queen can be moved to the other side of the main colony.

This encourages more flying bees to leave the queenless colony, further weakening it, and join the main colony which is strengthened and more able to collect nectar and make honey (5).

Weakening the queenless colony ensures that when a new queen is produced in this colony it is not likely to try to 'cast' with the new queen leaving the colony with half of the bees.

The beekeeper has placed the skep between branches at the top of the swarm and is now encouraging them to climb up into it. Smoking will encourage the massed bees to climb up into the waiting skep. A cardboard box can also make a temporary home for a swarm.

The Snelgrove method

Mr Louis Snelgrove used to live near Weston-Super-Mare and wrote a number of excellent books on practical beekeeping. His method requires a brood box with a set of frames and a special board called a Snelgrove board. This board looks like a cover board but has hinged openings on three sides both above and below the board (see illustration). The advantage over the Pagden method is that it requires no additional space in the apiary (the second brood box is above the old hive) and uses less equipment. It is more complex and some beekeepers find the manipulations a bit confusing.

The old brood box with all the frames is removed and replaced with a new box filled with frames (1). The queen is found in the old box and placed on the frame she is on in the centre of the new brood box (ensure that there are no queen cells on this frame) (2). The old hive is then rebuilt, with the queen excluder and any supers in their original place.

Instead of placing the crown board on the top of the last super the Snelgrove board is placed there with all the doors closed except for the upper one at the front of the hive (3).

Go through the old brood frames and remove any sealed queen cells (there should be none if the queen was present) and reduce the number of open queen cells to two (as with the Pagden method). The old brood box is then placed above the Snelgrove board and the original crown board and roof are then replaced above this second brood box.

The nurse bees tend to remain with the brood whereas the flying and foraging bees will return to the new brood box containing the queen.

After about a four days the little entrance on the Snelgrove board is closed and one at the side of the hive is opened (4) to provide the colony with a new entrance. At the same time the entrance below the Snelgrove board where the original entrance was located is opened. The new flying bees that have orientated to the upper box return to the original entrance and join the

Careful swarm prevention techniques may help avoid having a swarm in a difficult place. Taking a swarm like this could present a serious difficulty for an inexperienced beekeeper – the tree is tall and branches quite thin. Never be afraid to seek help or say 'no' in this sort of situation.

The Snelgrove method of swarm control

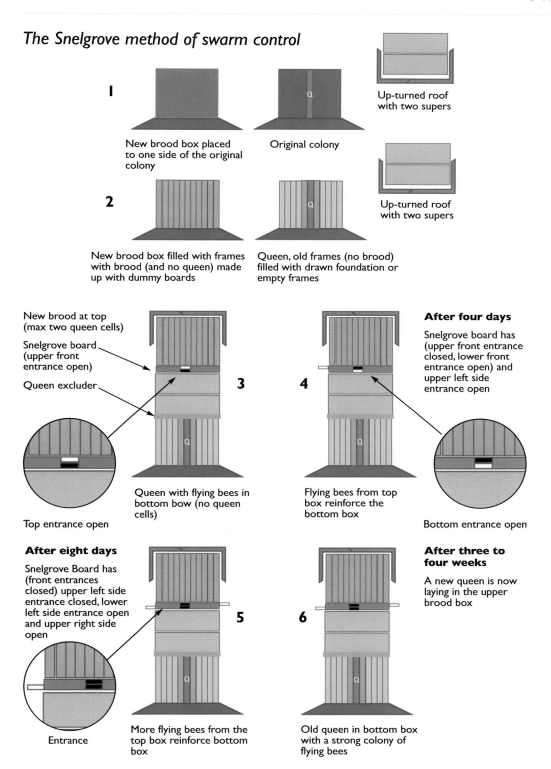

I

New brood box placed to one side of the original colony

Original colony

Up-turned roof with two supers

2

New brood box filled with frames with brood (and no queen) made up with dummy boards

Queen, old frames (no brood) filled with drawn foundation or empty frames

Up-turned roof with two supers

New brood at top (max two queen cells)

Snelgrove board (upper front entrance open)

Queen excluder

3

Queen with flying bees in bottom bow (no queen cells)

Top entrance open

After four days

Snelgrove board has (upper front entrance closed, lower front entrance open) and upper left side entrance open

4

Flying bees from top box reinforce the bottom box

Bottom entrance open

After eight days

Snelgrove Board has (front entrances closed) upper left side entrance closed, lower left side entrance open and upper right side open

5

More flying bees from the top box reinforce bottom box

Entrance

After three to four weeks

A new queen is now laying in the upper brood box

6

Old queen in bottom box with a strong colony of flying bees

The cast

After the prime swarm has left the hive if there is another queen cell present, the bees may decide to swarm again. This very much smaller swarm is know as a cast.

When checking the hive you must be careful not to leave a maturing queen cell tucked away under a pile of bees. If two swarm cells develop and two queens emerge almost at the same time you will surely loose a cast and the colony will be so depleted of bees that it is likely to fail.

Take care and be diligent with your regular checks – especially around swarming time!

The Snelgrove board with the front top entrance left open.

lower colony. The reason for doing this is that the flying bees will move down inside the hive and leave by the main entrance of the hive. After doing this a couple of times they will readjust to the lower entrance and no longer try to use the upper entrance. At this stage this upper entrance door can be closed to prevent predators gaining access this way (5).

This further reduces the population of bees the upper box and increases the number of bees in the bottom box. The result is that the lower colony has more bees to collect nectar and the upper box does not get congested and is unlikely to be able to produce a cast (a second or perhaps third but much smaller swarm).

Once the queen has hatched and is ready to go on mating flights, she will do this from the side entrance and will eventually start to lay eggs (6). At this stage the two colonies can be united to produce a strong colony with plenty of bees and a young queen once the swarming season has ended. Alternatively, the two colonies can be separated if they are established sufficiently to survive the following winter.

The Snelgrove method of swarm control sounds complicated with a number of little doors being opened and closed. Once it has been seen demonstrated by an experienced beekeeper it is clear that it is quite an efficient process that is ideally suited to an apiary where there is not much spare space.

The Demaree method

This was propounded by the American George Demaree in the 1880s. It was originally suggested as a method of swarm prevention but these days is sometimes used for swarm control. The method requires an additional brood box but very little else. Once again the original brood box is removed and replaced with a new brood box complete with frames. The queen is found and placed in the new brood box on the frame where she is located after ensuring there are no queen cells on this frame. The hive is then reassembled and the old brood box is placed above the supers and below the crown board. The arrangement is identical to the Snelgrove method without the Snelgrove board.

Over a day the bees will naturally rearrange themselves so that the brood frames above have sufficient bees to develop the brood and the rest stay below with the queen and support her and continue to forage.

At this stage the upper brood box can be isolated from the rest of the hive by inserting a board with a single entrance below the brood box. The upper box will raise another new queen.

Although Mr Demaree designed this as a method of swarm prevention, it is almost inevitable that once the brood is separated from the queen that the bees will start to produce queen cells in the upper box. It is necessary to go through this box and remove all the queen cells after about five days.

Quite often beekeepers will start their method of swarm control using the Demaree method and then insert a Snelgrove board after a couple of days, a combination of the best of both methods!

The Horsley method

This was first developed by a Yorkshire man, Arthur Horsley, after having seen the Snelgrove method demonstrated by Louis Snelgrove. He felt that the various manipulations were rather complex and required too many visits to the hive (OK if the apiary is in the garden but not so good if the apiary is some distance away). He designed a board which is like a crown board but there is a piece of queen excluder on the side and a slider connected to an opening and a plate. When the opener is opened allowing bees to fly in and out the plate covers the piece of queen excluder and prevents bees from moving above and below the plate.

The process starts as it does with the Snelgrove board, isolating the queen in a new brood box in the same position as the old brood box. The colony is then reassembled with the Horsley board above the upper super. The old brood box with the brood is placed above this and the crown board and roof replaced. At this stage the entrance in the Horsley board is kept closed and left this way for about two days. During this time the bees

Arthur Horsley

Arthur Horsley was a pioneering beekeeper from Yorkshire. Like Snelgrove, Horsley invented a way of isolating the queen in a new brood box and thus reduced the urge of the colony to swarm, by removing the queen and flying bees.

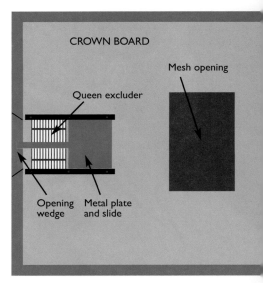

Drawing of the modified crown board developed by Arthur Horsley.

Newspaper choice

Note that it is not essential that you use a copy of the Financial Times to carry out the process of combining two colonies.

However, it is naturally easier to do this with a broadsheet newspaper.

are able to pass between the upper and lower boxes through the piece of queen excluder. The entrance is then opened, which simultaneously blocks the passage of bees between the colonies. The upper isolated colony then continues to produce a new queen as before.

It can be seen that all these method are very similar and operate by removing the brood from the queen and flying bees. The use of apparatus like the Snelgrove or Horsley boards are only there to make the manipulations simpler. Unfortunately they sound complex when described on paper but are easy to comprehend when the process is demonstrated.

Combining two colonies

Most methods of swarm control rely on separating the colony into two elements. So, if the beekeeper does not want to end up with an ever-increasing number of colonies, there comes a time when two colonies should be re-combined into a single hive. The simplest method is known as 'the newspaper' method and allows two colonies that would otherwise fight for supremacy in the hive to join together without any fuss.

The newspaper method

The idea is to bring the two colonies together until they are within a metre of each other. Move each of the hives by less than a metre every other day until they are close to each other. But be circumspect: if a hive is moved by more than this distance in a single move the returning bees may have difficulty recognising their hive and may try to enter another hive or fly aimlessly around until they find their hive.

Once the colonies are close enough, inspect the colony containing the queen that will not be required and remove the queen. Then open up the colony that is to remain on the final site, take off the supers and remove the queen excluder.

The beekeeper is cutting slots into the newspaper covering the brood box prior to the second brood box being placed over this paper.

Place a sheet of newspaper over the hive and replace the queen excluder. Using the hive tool make a few holes in the newspaper. Then place the second brood box on top with a second sheet of newspaper and queen excluder. Again slash the paper in a few places and place the supers from the first hive on top. If the second hive also had supers, separate these from the other supers with a queen excluder and another sheet of newspaper. However, this last manipulation – separating the two sets of supers – is rarely required. Beekeepers often ask if the queen should be in the lower or upper brood box. In fact it makes very little difference to the bees and the decision can be made on the basis of what is most convenient.

Over the next 24 hours the bees will chew their way through the newspaper and gradually, members of the two colonies will meet through the narrow gaps in the paper. The gradual meeting seems to remove the aggression from the bees and they will happily combine with the single queen.

Swarm collection

After a swarm has left the hive and settled in a nearby location, it is relatively easy to retrieve the bees and house them in a new hive. Problems occur if the swarm is allowed to remain long enough for the scout bees to find a new permanent home. For as soon as the swarm leaves the temporary resting place it is committed to the new home and will be very difficult to remove. The final resting place is invariably in a cavity and well protected. As soon at the swarm arrives the bees will start to produce wax comb to store nectar and to provide the queen with the opportunity to lay eggs. The new nest site then has resources that the bees will defend and be reluctant to leave.

The best time to catch and contain a swarm is as soon as possible after it has left the original hive.

Unless the swarm came from your own colonies and you were aware of its occurrence, you may receive a phone call from the police, someone from the local council, a worried member of the public or another beekeeper asking for your help. It is an unwritten law of beekeeping that all beekeepers should be able and prepared to collect a swarm if asked. This is because the swarm could become a nuisance and the bees will be better protected in a hive than in the wild. Before attempting to collect a

Swarm kit

Straw skeps and cardboard boxes are ideal swarm collecting equipment as they can be squashed into tight spaces.

Don't forget to take a small folding saw, a lighter or matches for the smoker and the all important queen cage – just in case you need it!

Using an upturned straw skep to collect a swarm that has settled in a nearby tree. If you don't have a skep then a cardboard box will do as well. Plus you can then keep all your swarm collecting kit in the box!

swarm on your own you should assist an experienced beekeeper so as to learn the procedure and be fully aware of all the safety issues involved.

When collecting a swarm it is sensible to wear fully protective clothing (a bee suit) even though bees in a swarm are normally very gentle and disinclined to sting. To collect a swarm you need the following equipment:

- Sheet to lay on the ground under the swarm
- Receptacle to catch and contain the swarm; a skep (straw basket for housing bees) or a cardboard box
- Secateurs
- Smoker
- Soft brush
- Hive tool

Bees beginning to settle down waiting for the scouts to lead them to their new home.

The bees will often fly to a low branch on a tree at about head height and, after a while, quietly cluster there (left). Should this happen, the bees are likely to be calm and the swarm will be easy to collect and remove to a safe site to be placed into a hive. To collect the swarm, place the box just under the swarm and give the branch a sharp shake. The bees will simply drop off the branch into the box.

Once the majority of the bees are in the box turn it upside down on the sheet and place a stone or something similar under the box to hold the base just off the sheet on one side. The bees will fly in and out of the box and eventually gather near the entrance and fan their Nasonov glands in order to waft a scent to nearby bees. This is an indication that they are content to stay in the box and that the queen is with them inside. If she is not there the bees will eventually abandon the box and fly off to find the queen, either where the swarm was located or nearby.

If at all possible, leave the box until twilight. By this time almost all the bees will have entered the box and there will be very few still flying about. Now the box is ready to be removed. Remove the stone, pull the sheet over the box and tie it to make sure the bees cannot escape. Once this has been done, you can carry the complete apparatus and bees away, in a car if necessary, to an apiary where the bees can be placed into a hive.

Of course it is not always as easy as this. Sometimes the bees decide to settle up in a tree a few metres or more off the ground. Collecting them may then entail using a ladder or a long pole with a skep attached. Sometimes they will decide to settle in the middle of a thick hedge or up a lamp post. While it is possible to extract bees from these situations, it is not easy. If you find a

swarm in these circumstances it is better to ask an experienced beekeeper with swarm-collecting experience who will be able to manage the safety aspects of this type of venture.

The hive tool is always useful for giving access to the bees and manipulating any frames you may have taken to collect the bees. The secateurs are used to remove small twigs and other vegetation to give better access to the swarm. Before doing some pruning you must obtain the owner's permission otherwise you could be in a lot of trouble! The soft brush is helpful if the bees are on a wall as they can be brushed into your receptacle. If this is done carefully the bees will come to no harm. Finally always take a smoker and have it ready to use before you start. The smoke can be used to persuade the bees to move if they are in a dense hedge. If the receptacle is placed on top of the hedge a bit of smoke will encourage the bees to move up into the receptacle. There is a chance that the bees might get distressed and attack you. Puffing smoke about you will help to reduce any attack.

Above: Collecting the swarm in a skep. The bees had inconveniently settled in a plastic composting bin.

Left: A swarm that had settled in a wheelie bin being hived in a spare five frame nucleus.

In this case the queen was popped into a traveling cage for safety. The skep was turned over, one side raised whilst the flying bees settled into their new and temporary straw home.

Hiving a swarm

Having got the swarm back to the apiary you need to put it into a hive. Usually the swarm is not too large and can be placed in a nucleus box that holds five frames. Two main methods can be used to do this. One is very attractive to watch but can take more time than the other. New beekeepers will delight in the first method but with experience may favour the second to save time.

Method 1

Fill the nucleus with five frames of wax foundation or drawn comb. It is usual to use four frames of foundation with the fifth of clean drawn comb if it is available. The reason for having one frame drawn out is that it is easier for the bees to hold onto a drawn frame and it gives the colony a ready made place to store nectar and for the queen to start laying.

A newly-hived swarm is in the right condition for drawing foundation into wax comb as this would have been the first task when the bees arrived at the new nest. Place the nucleus on a stand and put a flat board in front of the hive, sloping from the ground upwards until it reaches the entrance to the nucleus hive. Gently unwrap the sheet you used to wrap the swarm and place it over the board. Carefully turn the box upside down over

Top: Shaking bees off the composter roof into a straw skep that has been placed on a large piece of hessian sacking.

Right: A swarm streaming uphill on a piece of plywood, set at an angle between the hive entrance and the ground, into their new home. The queen was placed inside the hive and now the mass of workers are following her.

the sheet and give it a sharp tap. All the bees will drop out onto the sheet and appear somewhat confused.

Some of the bees will start to walk away from the pile and as they prefer to walk upward a few will eventually find the entrance to the nucleus box. After a few minutes these bees will enter the hive and decide that it will make an excellent new nest At this stage, they will come out and at the entrance, start to expose their Nasonov gland to attract the other bees. Some bees will join them and start to do the same thing. Within a few minutes all the bees in the swarm will sense the pheromone and walk towards the entrance. Sometimes the queen will be seen rushing across the other bees in her dash to enter the new hive. After a few moments, virtually all the bees will be inside the hive and you can remove the sheet and board. The movement of the bees as they enter the hive almost looks like a stream of bees 'flowing' into the hive and is a sight that all beekeepers find delightful. This process can take about 30 minutes from start to finish.

Method 2

Leave the hive empty and place it on the stand. As above, the five frames are left ready for use but for the moment are left beside the box. Leave the top off the hive and once you have removed the sheet from the box, hold the box carefully upside down over the nucleus box. Give the box a sharp tap and the swarm will fall inside the hive. Return the five frames to the hive, one by one and rest them gently on the pile of bees at the bottom of the box. These frames will gently sink though the mass of bees until they stop on the supports in the hive. At this stage, you should replace the roof and all the bees that did not go into the hive or subsequently flew off will circle about until they are called back to the hive by workers at the entrance exposing their Nasonov glands.

From start to finish this process will take about five minutes but the stragglers may take a further 15 minutes to find their new home.

Be ready for swarms

As your beekeeping skills progress, you will naturally want to increase the size of your apiary and swarm collecting can be a ready source of new stock.

Keep a box of swarm collecting equipment at the ready, so that when you receive a phone call for help, you will be ready to go and inspect the site where the swarm has settled.

If it's in a tricky position – leave well alone and ask a more experienced beekeeper to help you.

Remember to quarantine the newly hived swarm from any other hives for a few weeks in order to check there is no disease present with this free bounty.

Workers at the hive entrance with their abdomens in the air exposing their Nasonov glands and fanning their wings.

Bee and colony health

Honey bees have evolved to their present form over millions of years. They have developed defences against the predators and diseases that attack them and have become well adapted to their environment.

When allowed to live in their natural surroundings, honey bees can cope with most situations that may threaten the existence of a colony. It is quite natural for the number of colonies in an area to plummet when faced with an attack from a new form of a disease.

However, once the honey bee has developed a strain that is less susceptible to the disease, the numbers will rapidly return to a normal level. It may take hundreds of years but the new strain will be stronger and ready to face new threats. Similarly once honey bees become resistant to a pathogen, the pathogen will adapt to be able to prey on its host again. There develops a natural balance between the honey bee and its pathogens where both can co-exist without the destruction of one or the other.

As honey bee colonies live and breed in a confined space, it is inevitable that disease pathogens will build up in the nest. The natural response to and defence against this is for the colony to swarm, find a new nest and build new, clean honeycomb. Once the bees have gone, wax moths and other animals will enter the old nest and devour its content, removing any disease left in the comb. The old site will now be ready to be occupied by another swarm when the time is right.

Keeping honey bees in a hive means that the responsibility for ensuring disease does not build up inside the hive rests firmly with the beekeeper. To keep the bees healthy, the hive must be cleaned regularly and effectively.

The importance of hygiene

Research has shown that if disease pathogens do enter a hive they are eventually deposited in the wax comb. Larvae that are diseased will either quickly die or develop into weakened adults. Dead larvae are often removed from the nest by adult bees and the disease may be removed with them. This is not always the case and if diseased larvae are left in the nest the bodies will break down and release the disease into the cells. Should an infected larva survive to adulthood it will defecate in its cell as part of the pupation process. This releases disease into the cell and the resulting adult will continue to be a further source of disease. Once a cell has been used to raise a new bee it will be cleaned out and re-lined by young workers. If the cell is infected, the workers will pick up the disease on their mouthparts and continue to spread it though the nest and onto other bees.

Some of the diseases that affect honey bees are fungal and if these kill a larva, the fruiting bodies will produce fungal spores that will pervade the whole nest.

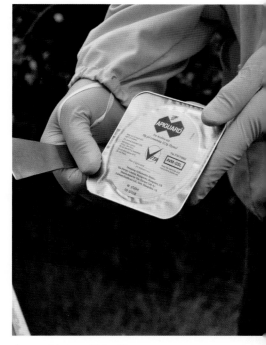

A tray of Apiguard® for *Varroa* treatment.

It is good practice to replace the wax comb in the brood chamber regularly and to destroy the old comb by fire to get rid of any disease left on it. A good general rule is that if a wax comb has been used by the colony to raise brood for two years it should be replaced. There are a number of methods for doing this. The three most common are:

Replacing comb

There's a school of thought that suggests that comb should be replaced annually.

The comb below is old and clearly needs replacing as it's now well past its best.

Old frames may be used and rewaxed with fresh comb after careful cleaning. However in an attempt to keep hives in the best possible condition and free of disease and pesticide residues it is probably best to change all combs and frames each year.

This can best be achieved using the Bailey frame change method as described in this chapter.

Comb removal

At the end of the season when the colony is preparing for winter, inspect the brood frames. Move the older (darker) ones that do not contain brood at the time to the side of the brood nest. In spring, when the colony is building rapidly, remove these dirty frames and replace them with new frames containing wax foundation.

This is a simple process but means that even the new frames will soon be covered with pathogens that have been transferred by the bees from the remaining older combs in the nest. The pathogens in the brood chamber itself will still be present but reduced in number. The approach is adequate but there are better ways of replacing old comb in the brood chamber that will further reduce the risk of retaining sources of disease.

Bailey frame change

The Bailey frame change management approach relies on having a second brood chamber and a set of frames ready with new foundation (1). The technique is best done in late April when the colony is building up rapidly and there is plenty of forage for the bees.

The second brood chamber will be clean and disease free. Scrape all the old wax and propolis from inside the original chamber and then use a blowtorch to scorch the inside. This will kill many disease pathogens and others will be embalmed in any traces of wax that remain in the chamber.

Place the brood box above the old brood box with a queen excluder between the two (2). If there is no flow of nectar, indicated by plenty of flowers in the vicinity and many bees flying to collect food, then the colony will need to be fed. The workers will eventually move to the new brood chamber and start to draw out the comb (3). Once two to three frames are fully drawn, go through the lower box and find the queen. Place her on a drawn frame in the centre of the upper box (4). Reassemble the hive with the entrance to the lower box facing backwards (turn through 180 degrees). Remove all frames in the lower box that do not contain any brood, replace the queen excluder and

The Bailey frame change method
(a strong colony)

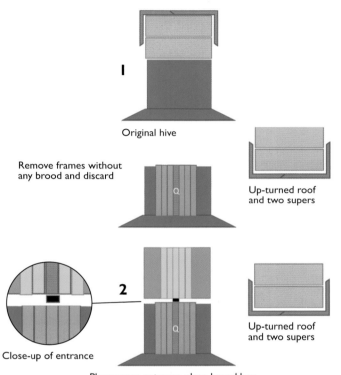

1

Original hive

Remove frames without any brood and discard

Up-turned roof and two supers

2

Close-up of entrance

Up-turned roof and two supers

Place a new entrance, clean brood box and frames of foundation **above** the queen excluder

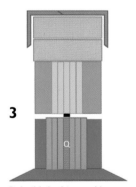

3

Rebuild the hive and leave for one week until new foundation is drawn out

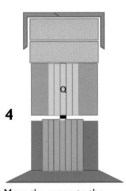

4

Move the queen to the upper brood box and check after 3 days for any queen cells in lower box and destroy them

5

After 3 weeks, remove lower brood box and destroy all frames in it. Add more foundation in the new brood box as needed

A garden apiary, with hives painted green and well concealed from the road by trees and shrubs.

A weak colony

If the colony is weak, perhaps suffering from nosemosis, it is unlikely that the bees will be able to leave the queen and draw wax foundation in the upper chamber (1).

Under these circumstances the queen can be moved up to the upper brood box and surrounded by frames of drawn comb (2).

Feeding the colony will encourage the queen to continue laying and the bees will migrate to the upper chamber (3 & 4).

After 3 weeks the lower box can be removed and the frames discarded (5).

Removing supers prior to inspecting the brood, here a 14x12 or deep brood box, that holds about 60,000 bees in full season.

The Bailey frame change method
(a weak colony)

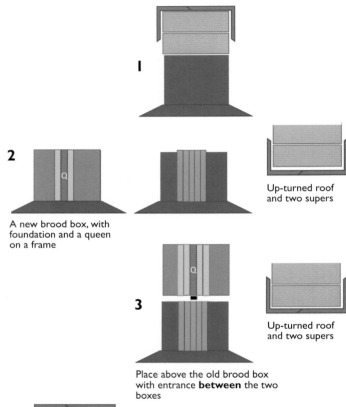

2

A new brood box, with foundation and a queen on a frame

Up-turned roof and two supers

3

Up-turned roof and two supers

Place above the old brood box with entrance **between** the two boxes

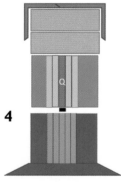

4

Replace the supers and wait for brood in lower box to hatch. Check the lower box for any queen cells

5

After 3 weeks, discard the lower box and fill the new brood box with foundation then place an eke (this is a spacer placed between two boxes to increase the gap between them) with an entrance facing forward above the queen excluder (it is possible to acquire a queen excluder incorporating a small entrance that will replace the need for an eke). Close up the hive.

The lower brood chamber will need to be inspected after about four days and any queen cells removed. The queen will continue to lay eggs on the new frames and brood will continue to hatch in the lower chamber. Once all the brood has hatched this (lower) brood box can be removed and the hive assembled as normal. All the old brood frames should be burnt to kill off any disease pathogens (5).

This is a better way of removing diseased frames but still allows some cross-contamination as the bees move from the lower chamber to the upper one and also from the frame bearing the queen.

Top: Inspecting the brood box.

Left: Demonstrating the Shook swarm method of frame changing. The beekeepers are replacing old comb, that may have been diseased, with a new set of frames and new wax comb.

Shook swarm

This method is probably the best approach for removing pathogens from the hive but also can be very stressful for the colony. It is most effective and successful if it is done in late

Top: Shaking old frames into the box half filled with fresh frames and comb.

Right: A pit dug specially to hold frames that need to be destroyed after an outbreak of European Foul Brood (EFB) disease.

spring (at the end of April). This method should not be used if there are queen cells with larvae in the hive unless you have first made sure the queen is still present.

In this method the old brood chamber is moved to one side. Place a queen excluder on the floor of the hive and a new, clean brood chamber on top of the hive. Put six frames of foundation in the new brood chamber (box), three at each end, to leave a large gap in the middle.

Then lift out each frame from the old brood chamber and shake all the bees into this gap. This is done for every frame, including the one containing the queen. The result is that all the bees are placed into the new box while all the brood, stores and other items are left in the old box. Then gently lower the remaining five frames of foundation onto the mound of bees and the reassemble the hive.

The colony must then be fed with large quantities of sugar syrup. It has lost all its stores and its brood and will need to rebuild the nest and care for new larvae. Placing the queen excluder on the floor prevents the queen from absconding with the other bees. After about two weeks you can remove the queen excluder as by that time the foundation will have been drawn and there will be new brood in the nest.

Destroy the old comb with all its stores and brood. This is a drastic process but evidence shows that when done correctly and at the right time, the colony will build faster than it would have done and that the outcome is a colony free from all disease that may have been in the hive. The only contamination is that left on the adult bees.

The beekeeper's personal equipment

The beekeeper is often considered to be the main means of transfer of infection from one colony of bees to another. When a diseased colony is inspected, it is inevitable that some infected material will find its way onto the bee suit, gloves or beekeeping tools. Whilst an experienced beekeeper may recognise the evidence of a diseased colony, newer beekeepers may find this hard to discern. When dealing with your own bees it is advisable that you clean all your clothing and beekeeping tools each time you visit the bees.

Bee suits
Bee suits are easily cleaned in a washing machine and should be washed after every visit to the apiary. The veil should not be washed at temperatures higher than 40°C. If you visit another apiary such as your local branch apiary or your mentor's apiary it is also a sensible precaution to wash your suit afterwards.

Gloves
Some beekeepers suggest that wearing gloves is a sign of a poor beekeeper; we believe that this is completely wrong. Our advice is to wear disposable gloves, made of either latex or nitrile. The latter is more suitable if you may have an allergy to latex, or alternatively you can wear rubber washing-up gloves. Washing-up gloves can be cleaned with warm water with washing soda added and can be worn a number of times, whereas disposable gloves are for use once only. However disposable gloves are thinner allowing greater dexterity and touch when manipulating the bees.

Many people advocate using bare hands, but if there is disease present in a hive it is more difficult to clean hands than it is to

Keeping frames infection-free

Old brood frames are always a potential source of infection.

Wax that is over two years old should be destroyed, whereas frames may be recycled.

First, cut out all old wax from the frames and sterilise the bare wood with a solution of washing soda (if you can be bothered).

In any case it's probably a good idea to buy in new frames on a regular basis.

Always work properly kitted out with a clean beesuit and a pair of disposable nitrile or similar gloves.

Cleaning your hive tools

A suitable container for cleaning your equipment can be an old 'water based' paint container or better still ask your local shop for one of their used ice cream tubs.

These useful tubs are often thrown out when the ice cream has been sold and make a superb and easily cleaned home for all your hive tools.

Don't forget to ask for a lid to complete your cleaning kit!

replace disposable gloves. Bare hands may also leave a smell in the hive that may disturb the bees. Furthermore, should you need to drive to your apiary, bare hands can get quite sticky when managing bees and this may come off onto surfaces in your car.

Finally some advise the use of leather gloves. These have the advantage that it is difficult for bees to sting your hands through the leather but the 'feel' through the gloves is reduced to such an extent that it is possible to crush your bees without realising you have done so. Leather gloves are very difficult to clean and so disease and alarm pheromones from the bees can build up to levels that can infect the bees or encourage bees to attack the gloves. For all these reasons we would recommend staying with washing-up gloves or disposable gloves at all times.

Hive tools and smokers

Hive tools and smokers are other possible sources for the transfer of disease from one colony or apiary to another. Hive tools and other items used to help manipulate the colony should be kept in a bucket containing about a five per cent solution of washing soda with a touch of washing-up liquid added. This will help to keep the tools clean and free of any pathogens. A stainless steel scourer is useful to rub off any bits of wax or propolis that remain on the tools.

Above: Carrying a bucket in which to put any brace comb found during inspection.

Right: Cleaning hive tools with washing soda solution.

The smoker barrel is usually free of pathogens as it is hot in use and should not come into contact with the internal components of the hive. However, the bellows do get contaminated, from the gloves, and should be cleaned regularly using a stainless steel scourer.

Pests

A number of pests may try to attack a honey bee nest, either when seeking shelter or food. When bees are kept in a hive the beekeeper must help them to protect themselves.
The common pests are as follows:

Wasps, hornets and other bees

By August, wasp nests begin to collapse and the adult wasps become increasingly starved so will look for any sweet liquid. The honey in a hive is attractive to wasps and they will attack weak colonies to obtain the honey stores and any larvae they can find. The attacks can be devastating and result in the destruction of the colony. Bees from strong colonies will also try to enter weaker colonies to steal the honey for their own hive. Once these attacks start they are very difficult to stop because the bees and wasps will indicate to others in their colony that there is food in the bee hive.

You can provide protection by ensuring that there are no holes that give access to the hive apart from the entrance. Reduce the entrance size so that it is no wider than two or three bees and about one bee high. This makes it easier for the guard bees to defend the colony. Hive manufacturers produce entrance blocks that reduce the entrance for these eventualities. If the wasp attacks are severe you can try to locate the wasp nest and destroy it. By this time the remnants of the old wasp nest are doomed and all the wasps, except the new queens, will die in the near future. Queen wasps will find a secluded spot (in an old mouse hole or under a roof) where they can hibernate until the following spring.

Hornets are not usually a problem to honey bee colonies in the UK but have been shown to be a problem in Europe and particularly Japan and the Far East. The Asian Hornet (*Vespa vetulina*)

The Asian Hornet

Beekeepers have been asked to report any sightings of the Asian Hornet, which at the time of writing has not been seen in the UK although it is in France.

We understand that over 500 deaths have been attributed to this pest in China each year.

In late summer wasps can become a nuisance raiding hives for honey and eating windfalls to obtain food.

Protecting hives

The hives below have been protected with plastic sheeting to prevent Green Woodpecker damage.

In addition mouse guards are fitted to the entrance of both the near hives. Around the farthest hive there is a wire mesh tied to an old tree stump and supported by metal stakes – again to prevent woodpecker damage and any possible incursion into the apiary from badgers.

The red 'card playing' symbols on these hives are said to help bees find the correct hive when returning after a flight away from their home.

has been shown to be a threat to honey bee colonies that have virtually no way of combating an attack from these large insects. Small entrance blocks will help but the best approach is to trap the queen hornets in spring before they have a chance to build a colony. Hornets tend to be gentle until their nest is attacked when they will mass attack the intruder. The sting they inflict can be very painful and dangerous so it is not advisable to try to remove or kill a hornets' nest. If one is found leave its destruction to professional pest control experts.

Mice

Wood mice, house mice and other mice will try to find a warm and dry place to hibernate in the winter. A beehive is an excellent substitute for a small hole in the ground. Once winter comes and the first frosts are imminent, mice will seek a nest. If a mouse enters the hive the bees will already be clustered for winter and will not reject the intruder. The mouse will make a straw nest at the base of the hive, chewing wax comb and any other materials to add to the nest. However the mouse will upset the bees and its presence and the disturbance it causes

A perforated, zinc coated mouse guard that is pinned across the front entrance to the hive. The bees can pass through the holes but not the mice.

Both plastic covered hives in the middle picture have similar mouse guards fitted under the red 'playing card' symbols.

may eventually result in the death of the colony or at least the destruction of some of the frames and a severe loss of bees.

Mice can be excluded from a hive by having an entrance that is small enough to prevent the animals getting inside. You can buy mouse excluders that go over the entrance and leave holes of about 8mm diameter.

An alternative is to ensure that the floor of the hive has an entrance that is only about 7mm deep. Proprietary mouse guards may cause damage to the wings of bees as they pass through and can also knock pollen off the pollen baskets when the bees are bringing pollen back to the hive. Protection at the entrance of the hive should be in place before the first frosts start; otherwise a mouse may be trapped inside the hive.

Woodpeckers

The Green Woodpecker (*Picus viridis*) normally feeds on ants, using its long, powerful beak to dig into the soil to find ants in a nest. As the ground gets colder and harder when winter arrives, the ants move deeper in the ground. Woodpeckers will then look for alternative food sources and a beehive becomes very attractive. A Green Woodpecker will bore a hole through the side of the hive and, using its long tongue, feed on bee larvae and honey.

It appears that Green Woodpeckers need to learn how to find food from a beehive and they are taught to do so by experienced members of their species. The result is that in some parts of the UK, woodpeckers are a continuing nuisance to beekeepers whereas elsewhere, Green Woodpeckers and beehives coexist with no damage to the bees.

You can protect a hive from Green Woodpeckers by placing a wire mesh cage over the hive leaving a gap between the hive and the cage so that woodpeckers cannot gain access to the hive. It is then left there from autumn to spring. An alternative is to place strips of plastic hanging down from the roof and over the sides of the hive. This also stops the woodpecker from gaining a hold on the side of the hive to penetrate the side. There is a picture of a hive protected this way on the right.

Badgers

Badgers may also be a nuisance to beekeepers. Should a badger come across a hive on one of its trails it will knock it over and

Temporary repairs to the brood box of a hive following damage by Green Woodpeckers (*Picus viridis*) who have pecked holes through the brood box.

Wire mesh fitted around the hive can now be seen more clearly than in the picture opposite. This prevents damage by Green Woodpeckers (*Picus viridis*).

Badgers and WBCs

Young badgers seem to enjoy playing with the sliding entrance bars of WBCs. It is best to fit a perforated mouse guard otherwise the cheeky little blighters will pull the sliders out or run off with them!

feed on the contents. An attack by a badger is rare and can easily be prevented by placing the hive on a concrete block and strapping the hive together along with the block. It is very unlikely that a badger will be able to topple the hive if it is fixed in this fashion and, even if the hive is toppled, it will not break apart. It is good advice to never put a bee hive near a badger trail.

Wax moth

There are two species of wax moth, the Greater and Lesser Wax Moth. Both have an important role in helping to use up all the wax and other debris in an abandoned bee nest and clearing away any disease in the comb. But wax moths can be a nuisance within the hive during the active season and may also attack frames of wax that have been put away over winter.

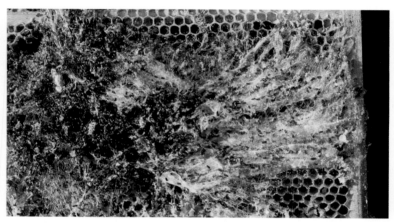

Wax Moth and right the damage to the frame it causes. Left undetected it takes very little time for a few Wax Moths to devastate an entire colony.

It is usually the Greater Wax Moth that develops in the brood nest of a colony. The larva will make tunnels through the nest underneath the cappings. In their progress they can damage larvae and leave an unsightly mess. Once fully-grown the larvae crawl off and chew a boat-shaped depression in any wooden part of the hive or frames where they will pupate into an adult moth.

When inspecting a frame, the presence of Greater Wax Moth can be detected by a trail just below the surface of the cappings. If the frame is tapped with the hive tool, the wax moth larva will often lift its head out of the trail and can then be removed from the hive. Any larvae and pupae should be destroyed whenever they are seen.

Lesser and Greater Wax Moth can also damage comb that has been stored away from the hive during winter. They are particularly active if the comb is stored in a warm location. You can destroy eggs, larvae and adult moths by fumigating the comb with 80 per cent acetic acid. Place the combs in the brood and super boxes and stack in a safe place, wrapped in a large polythene bag. Place a pad soaked in acetic acid (about 100ml per box) at the top of the stack and seal the bag. The stack should be left outside in a place where it will be cold but protected from high winds. Acetic acid is highly corrosive and will corrode any metal parts (like frame runners), which should be protected by a wipe of petroleum jelly. It will also destroy the structure of concrete so care should be taken not to get the acid on yourself or on any concrete.

Wax Moth

There are many useful leaflets published by the BBKA, amongst which is a leaflet about Wax Moth. A copy may be found on the BBKA web site.

Visit www.bbka.org.uk/files/ library/wax_moth_1020_ (data)_r2_1342860174.pdf) for more information.

An alternative treatment is to spray all wax with a diluted solution of a commercial caterpillar killer called B401. This is an active solution of BT, an effective and organic insecticide that is harmless to bees. Having sprayed the frames and allowed them to dry they can be stacked and stored in a cold dry location for winter.

Sheep and lambs can also be a nuisance to beekeepers, for as well as cattle they will rub against unprotected hives. It can be worth the investment to put up a secure fence around your hives if sheep and cattle are kept nearby.

Cattle

If hives are kept in a field alongside cattle they need to be protected from them. Hives are seen as useful rubbing posts and cattle will frequently knock over the hives. It is always advisable in these situations to discuss the risk with the farmer and put a stock fence round the area where the hives are located and top

A brand new hive painted in a sage green colour to help blend into the background is being conditioned in the apiary for a few weeks before any bees are installed.

Below: Healthy larvae with well capped brood cells together with honey stores.

the fence with two strands of barbed wire. This effectively keeps the cattle away and provides protection for the bees.

Vandals and theft

Although it seems strange, some people seem to get satisfaction from attacking a beehive. This is not normally a concern if the bees are kept in the garden but in a field, allotment or on a piece of rough land hives are often either knocked over or broken up. It may even be seen as bravery to attack a hive!

There is no foolproof protection from this sort of vandalism. Paint the hives in neutral colours to help ensure they are not easily spotted, and place them in positions that are not obvious from nearby paths and roads.

Believe it or not but beehives are subject to theft. The contents, honey and bees, are valuable and there are beekeepers who find it profitable to steal hives. It is sad to think that fellow bee-keepers are the felons but no one else has the equipment or wherewithal to remove a hive safely.

Beekeepers

Perhaps the greatest pest is actually the beekeeper! Beekeepers' fascination with bees and their lifecycles often mean they are tempted to inspect their colonies too frequently. When inspect-ing a hive it is possible for a beekeeper to damage some of the bees and even the queen. The only way to minimise this threat is to learn how to look after bees and treat any colony with respect.

Diseases
Brood diseases

Brood diseases affect the larvae in their cells. The most probable source of any brood disease is the adult bees that feed the larvae. These bees can pick up bacteria, fungi and viruses as they move around the hive and then pass them to the larva when feeding it. The result of brood disease is either that the affected larva will die before reaching adulthood or that it will be affected such that the resulting adult bee is weak and unable to support the colony.

In most brood diseases the general appearance of the sealed brood is an indicator that all is not well. If the sealed brood appears even, with few unoccupied cells, it is likely that the brood is healthy and all is well. Should there be many cells lacking cappings or brood at different ages adjacent to each other, giving an appearance of patchy (or 'pepper-pot') brood then there may well be a disease problem that needs further investigation.

As a new beekeeper you should never expect to be able to be able to identify and diagnose bee diseases on your own and should seek help whenever possible. At the early stages of bee-keeping it is only necessary to recognise that the brood looks healthy. If in any doubt, ask an experienced beekeeper for advice and if this is not available, then the advice of the local Appointed Bee Inspector (ABI) is always very helpful.

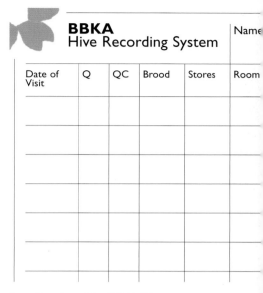

A small section of the BBKA's Hive Recording system.

Bacterial diseases

American Foulbrood (*Paenibacillus larvae*)

American Foulbrood (AFB) is a highly infectious disease of honey bees. Because of its virulence and ease of transmission it is a statutory requirement that it is notifiable to the Animal and Plant Health Agency (APHA) normally through an Appointed Bee Inspector (ABI). Fortunately the disease is rare (there are less than 100 cases each year in the UK) but if diagnosed in your colony, the colony has to be destroyed by burning, complete with the contents of the hive. The hive itself can be sterilised by scorching or the use of strong sterilising agents. Advice will be given by the ABI.

The standard field test for AFB is the 'rope' test using a matchstick to draw out (like a string) the larval remains as described in the text.

The bacteria form spores outside the larva, which can live for 30 years or more. Once the spores enter a larva along with its food they will develop and rapidly multiply. Within a few days the bacteria penetrates the gut wall of the larva and eventually the larva dies of septicaemia (blood poisoning). The larva usually dies once the cell has been sealed so the evidence to look for is that cappings have become dark, sunken, damp and perforated. Any cells that appear like this should be examined carefully.

You can attempt an initial field diagnosis by piercing the capping with a matchstick and then drawing out the larval matter. If it 'ropes' (draws out like a string) it is very likely that AFB is present and laboratory confirmation will be required. Severe cases of AFB will result in the dead larva 'melting' down in the cell and forming a hard scale at the base of the cell that is both highly infectious and impossible for the bees to remove. Lateral flow devices are available that will confirm the diagnosis but it is necessary to get the ABI to provide the final diagnosis.

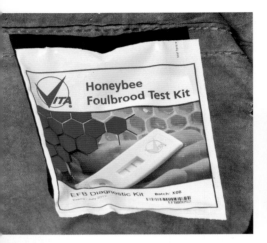

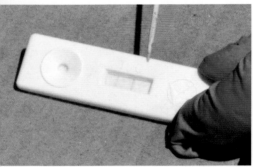

Top: The lateral flow device pack.

Above: These comprehensive test kits are available with full instructions on checking brood samples for EFB and AFB. If in any doubt, always seek further help from your Appointed Bee Inspector.

Right: European Foul Brood (*Melissococcus plutonius*) attacks mainly young larvae as seen in the picture opposite. EFB is caused by a non-spore forming bacterium.

The larva being pointed out is distorted and lying at an angle in the cell – a sure sign that something is not right. Larvae like this are not the normal ivory colour of a healthy larva.

European Foulbrood (*Melissococcus polutonius*)

European Foulbrood (EFB) is another highly infectious disease that is notifiable. Currently the number of cases in the UK is about 600 per annum, so it is still rare. Some beekeepers will keep bees for a lifetime without encountering this disease. Unlike AFB the bacterium obtains nutrition from food fed to the larva and, once it enters the larva, will compete with the larva for the brood food so that the larva is effectively deprived of food. The larva will appear contorted in its cell and normally dies before the cell is capped. Often the workers will remove these larvae and deposit them outside the hive but at busy times, they do not have time and dead and dying larvae can be seen in the comb. Should the food in the hive be plentiful, the larvae may get just about enough food to survive and pupate but the adult bee will still be weak and ultimately useless to the colony.

An alternative way to test for EFB is to spread a suspected larva out so that its gut can be seen through the body wall. The normal colour of the gut is creamy. If the larva is infected with EFB the gut fills with bacteria that are white in colour. The colour change is obvious and easily noted but once seen must still be confirmed by the ABI.

Fungal diseases
Chalk Brood (*Ascosphaera apis*)

Chalk Brood is a fungal infection that attacks larvae. The disease is passed on to the larvae in their brood food by the worker bees. The fungus develops rapidly and kills the larva. The larva then breaks down into a crumbly chalk-like pellet (called a 'mummy') in the cell and can easily be recognised. The mummies often fall out of the cells and can be seen on the floor of the hive. These mummies are loaded with spores that then spread over the frames in the hive. It is thought that chalk brood takes hold when the colony is weak and when levels of carbon dioxide rise above normal because the bees are not maintaining the correct conditions in the hive. Some strains of bees are more susceptible than others and re-queening can help. With bad cases it is advisable to replace all the brood combs to remove any sources of infection.

Stone Brood (*Aspergillus fumigatus*)

Stone Brood is another fungal infection of honey bee brood. Although it is rare it is of some concern because the fungus causing the infection is *Aspergillus flavus* that can infect man. The larva received the fungal spores in brood food fed by adults. The fungus multiplies rapidly and kills the larva, which then turns into a 'mummy.' Stone Brood can be distinguished from Chalk Brood as the mummies are very hard and can often look greenish when the fungus sporulates.

Chalk Brood 'mummies'

If you find black or grey Chalk Brood mummies in your hive the fungus will have produced spores.

These spores will have coated all the hive components and will almost certainly be a future source of reinfection.

If you have a more than just a few Chalk Brood mummies in the hive it is best to replace all the brood combs at an early stage.

An example of Chalk Brood. The mummified remains of a larva is clearly seen in this picture.

Cases are usually mild and clear when the colony is strong. In bad cases the colony should be treated in the same way as for Chalk Brood.

Viral diseases

A number of viruses affect both adult bees and larvae. Many have come to prominence since the arrival of varroa (see page 163) and are closely associated with the impact of the mite. The most common virus that affects larva is the Sac Brood virus and, while this has increased somewhat in recent years, it was present in many colonies before the arrival of Varroa.

Sac Brood virus (*Morator aetatulas*)

The virus affects the larva's ability to shed its last skin before pupating. The outcome is that the larva becomes trapped and eventually dies in the skin. When it dies it lies flat on the base of the cell with its head facing out. It resembles a Chinese slipper with the head end slightly turned up.

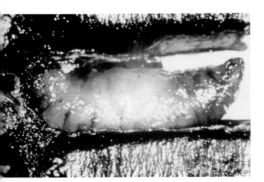

Examples of the Sac Brood virus showing the classic 'Chinese slipper' shape that larvae take on when infected with this virus.

There is no remedy for this condition and the advice is always to re-queen the colony at an appropriate time, as some strains of bees are less susceptible to the disease. It is rare for Sac Brood to have a major affect on the colony and it can sometimes disappear as workers can remove the infected larvae without becoming infected themselves.

Some infected larvae are able to pupate and emerge as adult bees. Interestingly these adults do not feed larvae and do not collect pollen. Their lifecycle is also accelerated so they die far earlier than uninfected bees.

This behavioural change helps to reduce the reinfection and spread of the disease but the diseased bees contribute very little to the colony.

Adult diseases

Adult diseases of honey bees are more difficult to spot, as a sick bee does not look any different from a healthy one. However there are other signs of adult disease and you can take a sample of bees to have them diagnosed by a beekeeping microscopist.

Nosema

Nosema (*Nosema apis and Nosema cerana*)

This is a fungal disease (called a microsporidium) of the gut of an adult bee. It is transmitted to the adult bee when it is cleaning the inside of the hive. The fungus develops in the gut lining (epithelium) and attacks the cells that assist with the digestion of pollen. The disease prevents the bee from absorbing proteins, and often-infected bees will develop diarrhoea. Affected adult bees will then defecate in the hive and, when other workers try to clear up the mess they become infected themselves.

Because the worker is debilitated by the disease it is less able to feed young larvae and less able to fly. Eventually it becomes useless to the hive. If nosema affects bees that live through winter, their lives will be shortened, and the colony is less able to survive through to spring.

The disease can be positively identified though a microscopic examination of a sample of 30 bees. Thirty bees are chosen as this gives a 95 per cent probability that the disease will be detected within this sample at a level of 10 per cent of infected bees. That is serious enough to require treatment.

The bees are killed by freezing, and then the abdomens are ground up and mixed with a drop of water. The mixture is inspected under a microscope at 400x magnification. Nosema shows up as small rice-like grains. There are a number of more experienced beekeeper that have microscopes and can help by doing this test.

Nosema can be controlled by an antibiotic called Fumagillin. However this is no longer licenced for use in the EU. There is at present no other medicine to treat Nosema but some products are said to help (Nozevit, Vita Feed green). Fumagillin was usually mixed with a sugar solution and then fed to the colony. Because the disease is spread through the hive by bees defecating inside the hive none of the above products are a long term cure. The fungal spores need to be removed from the hive and this is best achieved by cleaning the hive body and replacing all the brood combs.

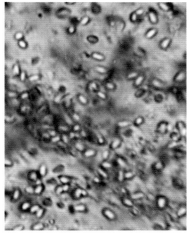

Nosema apis spores magnified x 400.

Above: Heavy dysentery stains at the hive entrance as a result of Nosema infection.

Acarine mites

When inspecting a bee for the signs of Acarine, once the head has been removed it is necessary to remove the 'collar' about the junction between the head and thorax.

This is a delicate process and requires practice. Once removed the tracheae at the start of the thorax can be seen.

Any discolouration from pearly white is an indication of an infestation by Acarine mites.

Sample of bees to be sent off for microscopic analysis.

If the colony is very weak then the best option is to kill all the bees, clean the brood box and other hive products and destroy all the brood frames. In less infected colonies it is possible to recover the colony by performing a slightly modified Bailey Frame change (as described earlier in this chapter on pages 145-146). The method is modified to take account of the colony weakness. The process starts in the same way but the queen is found and brought to the upper brood chamber on the frame she was already on. And the new entrance introduced at this stage. The reason for doing this is that the colony is probably too weak to build new comb in a brood box placed above the old brood box. Placing the queen in the upper brood chamber will encourage the bees to move there and work with her to extend the colony in this new area. Once the brood in the lower box has hatched out the hive is reassembled with only the new brood chamber and the old brood chamber is removed and the old frames destroyed.

Acarine (*Acarapis woodi*)

Acarine is a mite that lives in the first breathing hole (spiracle) in the thorax of the bee. The mite reproduces in the spiracle and clogs it up with mites and debris. There is no outward sign to suggest the bee is infested but it is believed that the life of the adult bee is shortened.

The infection is not considered serious in this country and it is thought that the treatments for varroa also affect acarine mites. The infection is worst when the colony is overcrowded because the adult mites move from an adult bee to infest a young bee no more than a few days old. Again the best way to treat the colony is to re-queen at an appropriate time as some strains are less affected than others.

The diagnosis can be confirmed by a beekeeping microscopist. Again a sample of 30 bees is required. After the bees have been killed, their heads can be removed to expose the breathing tubes just behind the head. If the bee is infected, the breathing tubes will be darkened, whereas a healthy bee will have clear breathing tubes.

Varroa (Varroa destructor)

Over the last 70 years or so, ever since the mite Varroa jumped species from the Asian honey bee to the European honey bee in the Far East, it has invaded colonies of honey bees and spread to most countries in the world. Varroa is probably the most dangerous pest that honey bees have to contend with. In the UK there are very few places where Varroa cannot be found in a colony of honey bees.

Interestingly, the Varroa mite (commonly *Varroa jacobsonii*) used to coexist with the Asian honey bee (*Apis cerana*). It was able to jump species onto our honey bee (*Apis mellifera*) sometime in the mid-20th century. The mite then evolved into a new subspecies (*Varroa destructor*) which in now unable to live with the Asian honey bee!

As it is not possible to eradicate Varroa, the approach that has been adopted is to keep the number of Varroa mites in a colony at a level where the colony does not show signs of weakening or damage, even if individual bees are affected. Varroa is a mite that affects both adult bees and larvae. It can live on adult bees and penetrates their casing to feed on their blood. It reproduces in the larval cells and feeds on the larvae. This debilitates the adult bees and larvae. What is worse is that in penetrating the skin, the mite transmits viruses that used only to be a minor problem for bees but now are major killers.

Honey bees have little defence against viruses and the best protection is to reduce the number of Varroa mites in a colony so that the viral infections do not get too serious. Research has indicated that if levels are kept below 1000 mites in a full sized colony, bees are able to thrive and show little effect from both Varroa and the viruses.

The Varroa mite is about 1.7mm wide and about 1mm long. It has eight legs and is quite mobile. Mites can be seen with the naked eye and are about the size of a pin-head.
The government web site: **www.nationalbeeunit.com/Public/ BeeDiseases/VarroaCalculator.cfm** shows how to determine the number of Varroa mites in a colony.

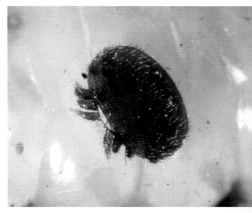

Top: Drone larvae exposed to show how voracious Varroa can be with young larvae. In a badly infected colony Varroa may be hidden away in large numbers within capped drone cells.

Bottom: Varroa mite on a honey bee larva

This is done either by counting the number of mites that fall off the bees onto a tray underneath the floor, or by uncapping drone brood and counting the number of larvae with mites on them. The effective monitoring of Varroa numbers is a technique that must be learned early in your career as a beekeeper.

There are many means by which to reduce Varroa mites in a colony and these should be used in combination to stop the mites becoming immune from any one treatment. The original treatment for Varroa was to use pyrethroid impregnated strips (Apistan® and Bayvarol®) inserted into the colony after the removal of the honey crop. This was originally very effective but continuous use year after year encouraged the development of a Varroa mite that is immune to pyrethroids. These products can only be used very infrequently now (no more than once every five years) and using a range of medicines in combination during the year has been found to be a better long-term approach.

Above: A pack of Apistan® strips

Right: Winter oxalic acid treatment of the hive brood frames.

In all cases, when using a medication it is important to follow the manufacturers' instructions and not to overdose or under dose the colony.

The simplest regime is as follows:

- Dust the bees with about 25g of icing sugar every time they are inspected from April through to August.
- In August or just after the honey has been removed from the hive; treat the bees with a Thymol (thyme oil) medication (i.e. Apiguard®) or a formic acid medication (i.e. MAQS®).
- Treat the colony with an oxalic acid medication (i.e. Oxuvar®) in mid-winter on a fine warm day.

This regime will keep the numbers of mites low and, as long as the instructions are followed carefully, will not harm the bees. As with all good medical practice you should monitor the number of mites before and after treatment to confirm that it has been effective and also to ensure that the use of medication is necessary.

There are many new management techniques and medicines being developed to combat Varroa. Researchers are trying to develop honey bees that are naturally able to manage Varroa without human intervention. This may eventually succeed but in the meantime beekeepers must look after their bees with great care and the minimal use of medicines and other substances.

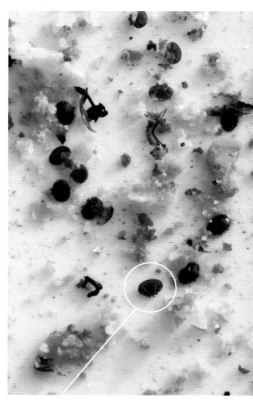

Above: Varroa mites on a hive floor tray showing how successful regular applications of oxalic acid and icing sugar can become.

Left: Drone with a deformed wing.

Top: Small Hive Beetle

Bottom: A Small Hive beetle trap made from a corrugated plastic material called Correx. This trap is inserted into the space between floor and brood box, where the beetles look for a suitable place to hide in the corrugations. The trap can be removed when the hive is opened and checked for beetles when doing an inspection.

Viruses associated with Varroa

The main virus associated with Varroa at present is the Deformed Wing Virus (DWV). The effect of this is that bees emerging from pupation are small with crumpled wings. They probably die within a few days and do not help to manage the colony. If bees are seen in this state then the varroa control is inadequate and should be improved. Slow Paralysis Virus (SPV) has also increased significantly since the arrival of Varroa as have a number of other less significant viruses. It appears that the reason for the increase is the mite's ability to penetrate the exoskeleton of a bee and such haemolymph. The virus is passed from the bee to the mite and then passed on to another bee. It has been suggested that whilst 'vectoring' the virus, the mite is also incubating the viral load causing more damage to the bees infected.

Small Hive Beetle (*Athena tumida*)

At the time of writing (March 2015) Small Hive Beetle (SHB) does not exist in the UK but has been found in Italy and may be on its way through Europe. It is a major threat to honey bees in the USA, Hawaii, Australia and Jamaica. The beetle can fly and will invade a colony some kilometres away from where it hatched and mated.

Once in the colony the female beetles lay eggs that develop into larvae that cluster and feed on bee larvae, pollen and nectar supplies. Once fully grown the larvae then leave the colony and pupate in the ground near the hive. Beetles then emerge and reinvade the colony or invade another colony and continue the cycle. The beetles are so destructive that they destroy the colony and all the contents of the hive.

Tropilaelaps (*Tropilaelaps spp.*)

Tropilaelaps was originally a pest of the Giant Honey Bee (*Apis dorsata*) but, like Varroa, another mite that has recently jumped species onto the European honey bee and could be even more devastating than Varroa. The mite is much smaller but can reproduce at a high rate, feeding on larvae in the nest. It is only able to live on larvae and may not be able to survive winter periods when little or no brood remains in the colony.

If Small Hive Beetle or *Tropilaelaps* arrive in the UK the first line of defense will be to attempt eradication. Should the beetle or mite establish in this country medicines will become available to control numbers along with traps to prevent beetles getting a hold in the colony. Should SHB be found then the colony infected will be destroyed and the surrounding ground drenched in chemicals that will kill beetle larvae in the ground. This work will be supervised by Appointed Bee Inspectors.

Notifiable diseases and pests

Because certain diseases can be transmitted easily from one apiary to another, the UK Government, in cooperation with the

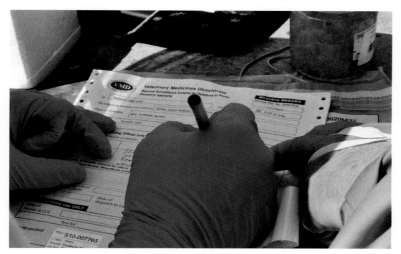

European Commission, has declared a number of pests and disease as notifiable. This means that should they occur in your apiary they must be reported to the Appointed Bee Inspector. These diseases are:

- American Foulbrood (AFB)
- European Foulbrood (EFB)
- Small Hive Beetle (SHB) and
- *Tropilaelaps spp.*

Top: An Appointed Bee Inspector (ABI) at work checking frames where disease is suspected.

Centre: Filling in the Veterinary Medicines Directorate (VMD) form.

Statutory requirements

If any of the notifiable diseases are suspected it is the responsibility of the beekeeper to contact the Appointed Bee Inspector (ABI), a civil servant, who will come and confirm the presence of the disease.

ABIs have the right to inspect any colony of bees to ascertain that it is free of disease and most cases of AFB and EFB are detected by ABIs doing routine inspections. If the ABI confirms a notifiable disease then a discussion with the beekeeper will resolve what treatment is required and whether the colony needs to be destroyed.

The inspector will help the beekeeper through the process and most beekeepers welcome their visits.

Top: Stack of hive parts being flamed to sterilise them before reuse.

Centre: The burning pit, where all frames will be destroyed from any hive with EFB disease.

The government runs a voluntary registration service for bee-keepers and their hives. This service is very important and should a disease break out near your apiary the Inspector will visit all local apiaries to find out the source of the disease.

Veterinary medicines requirements

Honey bees are considered to be food producing animals and are therefore subject to the medicine laws that apply to all other food producing animals (such as chickens or cows) All beekeepers are required by law to maintain records of all medication supplied to the bees. The records should contain the date and dose rate of the medicine together with the batch number of the particular product. No medicines should be applied outside the expiry date. The records should also be kept for at least five years.

The regulations may seem over stringent for an animal that collects food and then processes it by reducing the water content and adding a couple of enzymes, however the law is the law and the practice of recording what has been applied to the bees is a fair way to ensure that medicines are used responsibly. As with all other medicines it is very important to follow the manufacturers' instructions on dose rate and application method and frequency.

More details can be found on the National Bee Unit website – (www.nationalbeeunit.com/index.cfm?pageid=204).

British Bee Veterinary Association

Recently a number of vets have established the British Bee Veterinary Association (BBVA) comprising vets that are interested in bees and beekeeping.

It is hoped that this will provide a nucleus of vets across the country that are able to give advice on bee diseases and methods of treatment.

Bee Medicines Record Card

Beekeepers Name:
Address:

Product:		Supplier:	
Purchase Date	Quantity Purchased	Batch Number	Date, quantity an
Colony Identity	Date Administered	Quantity Used	Withdrawal Peri

Product:		Supplier:	
Purchase Date	Quantity Purchased	Batch Number	Date, quantity an
Colony Identity	Date Administered	Quantity Used	Withdrawal Peri

Above: An example of the approved VMD card available on-line from BBKA.

Left: Cleaning up after an outbreak of EFB infection.

Plants and hive products

Plants and bees have evolved together over many millions of years. Many need the other to evolve. Flowers have developed a number of features that make them attractive to bees. Cross-pollination has advantages because it increases the genetic diversity of the species and allows plants to respond to changing climatic conditions and spread their range.

Most flowering plants rely on attracting an animal by giving a reward of nectar. The flower will then transfer pollen on the animal's body so that it can travel to another flower which it pollinates by brushing against the stigma. This receives the pollen and starts the process of fertilisation. Other plants have developed other strategies to get pollen to travel from one plant to another of the same species. For example, grasses use the wind to scatter pollen to other grass plants. Some eventually land on a receptive stigma to start the fertilisation process.

Communication between plants and bees

Many insects are important as pollinators but bees are one of the most important worldwide, especially for agricultural crops. Honey bees are considered to contribute up to 50 per cent of the pollination of agricultural crops in the UK with the majority of the rest coming from bumblebees and solitary bees. Bees need pollen to provide them and their larvae with protein. In collecting pollen for food, they will also distribute it to aid plant fertilisation. Some species of bees and other insects have developed specialised strategies for collecting pollen from specific flowers and the continued success of many plants relies on the existence of these specialised pollinators.

Honey bees, on the other hand, are generalists that have learned how to get rewards from many different plants. In some cases this is simply to collect pollen for the colony and at other times it is to collect nectar with the distribution of pollen becoming an essential side effect.

Signs of pollination

Bees and flowers communicate in a very particular way. Bees can see ultraviolet light and if a flower is viewed in this part of the spectrum its petals will often have guidelines to help the bee find the nectaries and pollinate the flower. Some flowers can even indicate to bees that they have been pollinated and their nectaries are empty. A good example is white clover where once the floret has been pollinated it turns from white to grey and droops downwards. The central eye in a forget-me-not will turn from yellow to light grey once the flower has been pollinated. Other plants will drop their petals. These signs save a visiting bee from climbing inside the flower for no reward and no benefit to the flower.

Top: A clover flower that is showing signs of successful pollination around the base of the flower.

Bottom: Apple blossom awaiting a honey bee for successful pollination.

Scout bees are those that seek out new sources of pollen and nectar. They then return to the colony and dance to tell other bees where to find the new source. Other workers later visit the site to collect food for the colony. Bees will leave signs on a flower to give other bees information about the state of the nectaries. Should there be nectar in the plant the bees will mark the flower with the scent from their Nasonov gland to attract other bees to the flower. However if the nectaries are empty after a bee has finished feeding it can mark the flower with the scent from its mandible and this will act as a repellent to others. This scent is relatively short-lived, so once the flower has produced more nectar the scent will have dispersed and the flower will become attractive to other bees once more.

It has been shown that when bees settle on a flower the electro-static charge built up during flight is discharged through the flower. Another bee visiting the flower soon afterwards will be aware of the disruption in the flower's charge and be aware that the flower has been visited by another bee recently. The second bee will not waste effort trying to extract nectar and pollen from the flower.

Nectar and pollen collection

Nectar

If the bee is collecting nectar, it will gently land on the flower and then move in to the nectaries, which are usually found at the base of the petals, and gently drink up the pollen. These plants have anthers that are placed in such a way that, as the bee gains access to the nectaries, it brushes against the anthers and

Top: Searching for nectar in an Abelia flower in late summer.

Centre: Bees tending a full frame of liquid honey.

picks up pollen. When the bee goes to the next flower of the same species some of the pollen will be brushed off onto the stigma to pollinate the flower.

Nectar is drawn up through the proboscis (mouth tube) of the bee and passed into the honey stomach in front of the main stomach. The bee will visit a number of flowers until its honey stomach is full: it can hold about 40mg of nectar. Once it is full the bee will fly back to the colony.

During this time enzymes made in the hypopharyngeal gland will be added to the nectar. These enzymes are sucrase, which converts sucrose into the simple sugars glucose and fructose, and glucose oxidase, which in the presence of glucose and water will produce hydrogen peroxide, a powerful antiseptic.

Antiseptic properties of honey

It is the enzymes in honey that are mainly responsible for the healing properties of honey.

When applied topically (on the surface) on a wound the enzyme glucose oxidase generates hydrogen peroxide against the wound and helps to kill any bacteria in the wound.

At the same time the honey has a natural scouring action which help to clean any debris from the wound.

Top: Bees cleaning up some spilled honey.

Centre: Worker bees transferring nectar, known as trophalaxis.

Once the bee returns to the colony she will be met by a 'house' bee who will take the food from the forager by extending her proboscis into the mouthparts of the forager and eliciting the forager to regurgitate the nectar.

The house bee will then 'strop' the nectar. The worker repeatedly regurgitates the nectar onto its mandible and then swallows it again. The water content of the nectar is gradually reduced and water vapour evaporates from the droplet.

Eventually, when the water content drops below about 50 per cent the house bee will take the nectar to an empty cell and regurgitate the nectar into the top of the cell.

Some of the nectar brought back will be used immediately to feed the adult bees. Some will also be used to supplement the feed for the larvae. For immediate feeding the water concentration should be about 50 per cent to achieve the most efficient metabolism.

Production of honey

Droplets of nectar hanging in the cells above the brood nest are exposed to a warm and dry atmosphere. Water evaporates from the nectar and the solution becomes more concentrated. Sucrose, normal cane or beet sugar will only stay liquid until the sugar content reaches about 60 per cent, however a mixture of glucose and fructose will stay liquid when the sugar content is above 80 per cent. This is the reason that the enzymes added by the bees to the nectar are so important.

Furthermore the glucose oxidase in the honey will stop any natural yeasts present in the nectar from turning nectar into alcohol while the sugar content increases. Once the sugar content rises above 80 per cent the sugar concentration will be high enough to stop any yeast from fermenting the sugar solution. At this stage the bees will place a wax capping over the liquid leaving a small air gap. The honey has now been made from nectar and will remain stable under the cappings.

Balancing water and carbohydrate

A honey bee colony requires the carbohydrate in nectar for energy and the water for air conditioning. The 'house' bees are aware of the state of the hive and its needs. In periods of hot weather the inside of the hive can be warmer than is desirable and house bees will need water to evaporate to cool the hive. Under these conditions the house bees will empty the loads of foragers that come back with nectar low in carbohydrates and may even prefer it if the foragers bring back just water. These bees are often called 'tanker bees' as all they carry is water. Foragers learn from the house bees the needs of the hive by the

Top: Inspecting a frame for brood and sealed honey.

Bottom: Bees drinking some spilled unripe honey on a hive roof top.

rate at which they are unloaded and will adapt their foraging activities to meet the nutritional needs of the colony.

At other times the colony will need carbohydrates more than water. The house bees can regulate the income to the hive. If a forager comes back with nectar that is low in carbohydrate the house bees will be slow to unload the forager as a sign that the load is not what is required. This is an indicator to the forager that it needs to find a better source of nectar. Foragers that come back with higher concentrations of carbohydrate will be emptied quickly and can return to forage.

The messages transmitted from the house bees to foraging bees are quickly learned to ensure that the income to the hive is adjusted to the optimal ratio of water to carbohydrate based on the available nectar and water supplies in the area.

Pollen

If you watch bees collecting pollen from flowers it is clear that a very specific process is taking place. As bees fly through the air, their bodies become negatively charged with static electricity, much like clouds become charged to produce lightning. When the bee arrives at a flower it knocks pollen from the anthers, the static charge attracts the pollen particle and they stick in the branched hairs covering the bee's body.

Eventually the bee will become covered with pollen. At this stage it flies off the flower and hovers. It then regurgitates a bit of nectar and, using its front legs, wipes the pollen off the head. The middle and hind legs collect the pollen from the front legs and wipe the pollen of the thorax and abdomen. Eventually a sticky mass of pollen will be stuck to the brushes on the inside of the rear legs.

Pollen baskets

The bee then rubs its rear legs together and this moves the pollen mass upwards towards the joint on the rear legs. The joint is highly adapted to contain a press and backward facing hairs (see the diagram on page 12). The bee bends its legs and compresses the ball of pollen forcing it backwards and upwards.

Top: Collecting nectar from a *Ceanothus* flower.

Bottom: Worker bee returning with a full load of yellow pollen in her baskets.

On the outside of the legs the next section of the leg bears a 'pollen basket'. This is made from hairs that arch to form a basket with an opening at the bottom. Inside the basket there is a single strong hair. The movement of the presses in the leg joints forces the pollen onto this single hair. The pollen mass remains here while the bee continues to collect pollen or flies back to the colony. The single hair acts like the spike on the front of tractors that is used to carry bales of hay.

Depositing pollen in the hive

When its pollen baskets are full, the bee will fly back to the colony and, unlike in nectar collection, where it is helped by a house bee, the forager will enter the colony and find an empty or partially-filled cell and deposit the pollen directly into the cell. The cells used to store pollen are close to the larvae so that

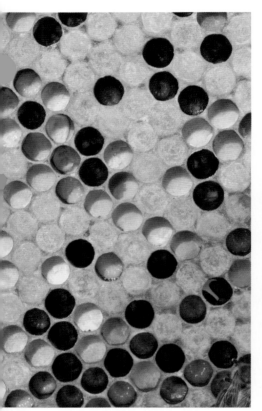

Top: Pollen stored next to cells containing larvae. A teaspoonful of local pollen taken twice a day is thought to help prevent hayfever.

Middle: Brood and food. Pollen close to the sealed cells and capped honey in a arc across the frame above the pollen. This would be typical of a good frame of growing pupae and stores.

the bees feeding the larvae do not have far to go to collect it. At certain times there is more pollen available than required by the colony. This is stored in cells further away from the brood and, when a cell is full of pollen the workers will mix a bit of nectar and enzymes with it and pack the pollen down. This does denature the pollen but prevents it from rotting away. The result is often called 'bee bread' and provides an emergency supply of pollen when none is available from outside.

Local forage

Bees remain healthier if they are able to collect pollen from a variety of flowers. Amino acids are metabolised from pollen by the bees but no one species of flower can provide the complete range of amino acids required by honey bees. If bees are restricted to one species of flower, such as oil seed rape (*Brassica napus*) they will eventually suffer. It is always difficult to make a significant difference to the range of flowers that your colony can find, as the bees will range over about 12sq km and, unless you are a farmer or have a massive garden the influence you can bring to bear on them is small. However every bit helps.

Try to locate your bees in an area where there is a variety of flora and where flowers can be found for most of the year. Honey bees often do better in urban areas where a range of small local gardens can provide a great variety of flowers and the flowering season is artificially extended. Similarly bees kept near unimproved fields, hedges and woodland can also find a good variety of pollen to enjoy.

It is always a good idea to grow plants that are attractive to pollinators in your garden as this will not only attract your own bees but also other pollinators into the garden so that you can

A selection of garden and native wild flowers help to provide a bees' five-a-day requirements.

enjoy watching them and helping all pollinators to survive. Try to encourage many of your neighbours to do the same. The most appropriate plants are those with open flowers and the more old-fashioned varieties. Select plants that will give an extended flowering period. If you have room, plant some trees that produce flowers that are single and open. Sallows and ornamental shrubs such as viburnum are always useful.

Some beekeepers move their bees from one crop to another as the season develops and provide a good pollination service as well as giving their bees plenty of pollen. This can result in large honey crops but does require a lot of lifting and carrying. The approach allows the beekeeper to collect virtually mono-floral crops. One that is particularly prized is heather honey from Common Heather or Ling Heather (*Calluna vulgaris*). This kind of honey requires additional equipment to extract it, as it is thixotropic (gel-like when still but goes liquid when stirred). Heather honey has a distinctive flavour that is liked by some but considered too strong by others.

When starting beekeeping it is better to learn the basics of bee-keeping before trying to get involved in the kind of migratory beekeeping outlined above.

Wax and propolis
Beeswax

Honey is not the only product of the hive that humans have found useful over the years. At one time beeswax was highly prized as it burns without smoke and no strong smell. When the only alternative was a smelly and smoky tallow candle, monks kept bees as much for the wax they produced as for their honey.

Beeswax is produced from glands on the ventral sides of the abdomen. At first it is almost clear but after it has been in the hive for some time it picks up pigments and stains from other material and gradually becomes brown or black. It is often stained with pollen and takes on a bright yellow colour.

If wax is to be used for making candles or blocks for sale or

Top: A large field of rape (*Brassica napus*) in full flower.

Bottom: Reclaimed wax tablets. The result of old super frames and any wax scrapings melted down in a steamer.

display it should be clean and a light colour. The most prized wax is that used by the bees to cap honey as this normally spends a short time in the hive and is the cleanest form. The worst wax comes from the comb used to rear brood. The larval cases stain the wax and it is virtually impossible to remove the discolouration. To prepare wax, the comb and cappings should be washed in soft water or rainwater to remove traces of honey and other soluble contaminants. It is then gently melted over a flameless heat (an electric hob) using a bain marie so that the heat comes via the water rather than being directed at the wax. The liquid wax is then poured through a fine muslin mesh or sheet of white linen to remove any solid matter. The liquid is poured into a mould and allowed to cool slowly. The outcome should be a block of wax the colour of primroses with no apparent cracks. If the wax is to be used to make a candle it can be poured into a silicone mould designed for the purpose.

Many beekeepers will produce blocks from their surplus wax and then exchange these for wax foundation from beekeeping suppliers. In this way a lot of wax gets recycled back to the bees.

Propolis
Propolis is a sticky substance that bees collect from buds on plants. It is used to polish the cells in which the eggs are laid prior to the queen visiting the cell. Propolis seals in any detritus

Care with wax

Wax burns at a high temperature and can easily be set alight if heated over an open flame.

Wax should be stored in the dark and away from any source of heat.

When processing wax it is advisable to heat it gently and keep the temperature fairly low (just above melting point).

Heating wax in a water bath is ideal as this limits the temperature to 100° C.

Top: Worker bee bringing back propolis.

Left: A frame corner covered with propolis or bee 'glue'.

left in the cell by the last bee that hatched there and has antiseptic properties that prevent some diseases from attacking the larvae. Propolis is also used by the bees to fill small gaps in the

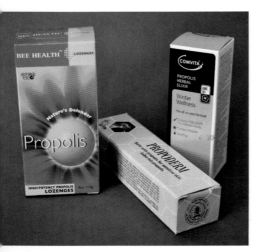

Top: A selection of retail propolis packs.

Bottom: Royal jelly is the creamy white substance in the bottom half of the queen cell.

hive. It may be used either to fill up holes to the exterior or to seal gaps where the bees cannot gain access but which might harbour smaller insects. Should a large creature such as a mouse or slug enter the hive and die, the bees will cover the offending object with propolis to seal it and prevent the deteriorating object from contaminating the hive.

Propolis can be used to produce an antiseptic tincture and is said to be helpful for sore throats. Some beekeepers mix propolis with an alcoholic spirit to make a gargle. There is a market for supplying propolis in bulk to health product manufacturers but the price paid is so low that most beekeepers do not bother to collect and sell surplus propolis.

Other products of the hive

Royal jelly (the food fed to queen larvae) is considered by some to be beneficial to humans and is collected by beekeepers with many hives to sell to health food suppliers. The majority of royal jelly that can be bought comes from China. Beekeepers in the UK do not have sufficient surplus to make sales a viable proposition.

Bee venom is also thought to have some medical benefits but its use is definitely not a part of western medicine. Dried venom is very dangerous and if breathed in can cause an anaphylactic reaction.

In some circumstances, bees are supplied to sting a person who believes this may improve their condition, whatever that may be. Supplying bees for such a purpose can be considered as an assault so beekeepers are strongly advised not to engage in this practice. The author was once very sensitive to bees' stings. A consultant observed him in hospital having made a bee sting him. The doctor believed that the reaction to a bee sting could be minimised by having a course of bee stings. In this case it worked, but was done with hospital supervision in case there was a serious reaction to the sting.

Extracting and processing honey

Honey for human use is removed from the hive in capped

frames from the supers. It can be taken from other places (frames in the brood chamber or wild comb) but this is not normally as pure as honey from the supers.

Removing frames for extraction

The frames are removed from the hive once the bees have been removed from the frame using one of the following approaches:

- Placing a clearing board under the super to be removed. The clearing board has a mechanism that allows the bees to go down to the brood but prevents them from returning to the super. The mechanism can either be a physical device that only allows bees to pass in one direction, (e.g. Porter bee escape) or a channel

that bees can pass down to the brood with comparative ease but where the obvious route back to the super is far less obvious to the bees (e.g. Canadian clearer board).

- Removing each frame and gently brushing the bees off the frame before placing the cleared frame into another super away from the hive.

There are many other ways to remove the bees but these two options are suitable for new beekeepers as they are relatively safe for the bees and fairly foolproof.

Oil seed rape

When clearing bees from a super of oil seed rape there is a danger that the honey will crystallise once the temperature of the super drops.

It is advisable to clear the bees from the super quickly (by brushing bees off frames or using a blower) and then extract the honey before it has a chance to crystallise.

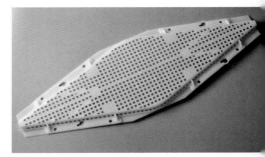

Above: A Canadian style clearer board. The plastic clearer device will be fitted to the underside of the crown board.

Centre: Inspecting a super of stored honey. This one has unripe cells that are not yet capped over, so it will put back into the super to finish off.

The supers are then taken away from the hive to an extracting room where the bees can be kept out and the area can be kept clean for the processing of a food. Many beekeepers use the kitchen or utility room where the surfaces have been cleaned and wiped with an antibacterial spray and there is no danger of any contaminants dropping from the walls and ceiling into the honey. It is important to keep all animals out of the room whilst honey is exposed and ensure that flying insects do not have access to the room.

You should wear clean clothes and cover over with a laboratory coat or something similar. Remember to wear a hair net to stop any of your hair dropping into the honey.

Uncapping

The frames are then taken from the super one by one and the cappings removed using a sharp knife or 'uncapping fork'. Special knives can be bought for uncapping. These have an electrical element embedded which enables the knife to stay warm. A sharp knife kept in warm water and wiped dry before use is a good substitute.

Extracting the honey

Once the cappings have been removed, the honey can be extracted from the frame by placing the frame in an extractor. This is a device that acts like a spin dryer. The frames are spun round and the honey flies out of the cells and runs down the walls of the extractor. The honey is then allowed to flow out of the extractor through a coarse filter that removes any lumps of wax or other debris. The honey is allowed to stand for about 12 hours so that air bubbles rise to the surface. It can then either be bottled, fine filtered, or stored in bulk containers for bottling at a later date.

Top: Removing wax cappings with a heated knife.

There are many designs of extractor especially designed for removing honey from wax comb with varying number of hold-ers for the frames. Most beekeeping associations have one that their members can hire for short periods. If the quantity of honey is not large the beekeeper may decide to simply crush the wax holding the honey and allow it to drip into a suitable container.

There are other ways honey can be removed from a frame, especially if there are only a few frames to deal with. The wax and honey can be cut away from the wax foundation using a special scraper. The resulting mixture of liquid honey and pieces of wax can then be filtered to remove the wax. Another way is to cut out all the wax from the frame and then put the wax and honey mixture into a fine cloth and press this to extract the honey from the wax.

On occasions the honey in the comb will solidify and become impossible to remove using an extractor. Beekeeping associations and suppliers can provide heated trays (often called Pratley trays) that will melt the wax and liquefy the honey. The mixture is then allowed to cool so that the wax solidifies and floats on top of the honey. Pratley trays operate at about 80°C and, as long as the honey does not remain at this temperature for more that about 15 minutes it will not be damaged significantly. However honey exposed to prolonged heating will be damaged in that the enzymes will be destroyed and the sugars in the honey will be converted into complex structures somewhat like those of toffee. The outcome is that the resultant material can no longer be called honey by law and is also not safe to feed to bees.

Storing

Honey is best kept in bulk. One of the most convenient ways is to pour the honey, immediately after extraction, into a 15kg food grade plastic bucket with a sealed lid. The bucket should be filled as full as possible to reduce any air and then stored in a cool dry place (about 10°C or lower). In the bucket the honey will solidify rapidly and become stable. As long as the water content in the honey is within the acceptable range (less than 20 per cent) honey can be kept this way almost indefinitely. It is advisable to measure the water content of the honey using a honey refractometer.

Top: Frames of exhibition grade sealed honey displayed at a honey show.

Bottom: Tub of solidified honey which will stay like this for many years.

When the honey is required for use the bucket and its contents need to be warmed gently to about 40°C in a water bath or heating cabinet for about 24 hours. This will return the honey to its liquid state so that it can be further processed. Do not exceed the temperature or time limits as this will also damage the honey. If these figures are greatly exceeded 'HMF' (hydroxymethylfurfuraldehyde) will increase and the statutory limit on this below 40 parts per million for the product to be called honey. So overheating can both damage the honey and remove the legal right to call it honey!

Filtering

The coarse filter used to remove wax and other debris from the honey when extracting it from the frames will still leave particles of other materials in the honey including pollen. Pollen is an important constituent of honey and should never be removed completely. However, if there is a lot of particulate matter in the honey this will encourage it to crystallise quickly and go solid.

If the honey is to stay liquid for some time it must come from the right floral source and should also be fine filtered to remove all but the natural pollen in it. The filters normally used are cloth mesh of either 100 or 200 mesh that can be purchased from beekeeping suppliers. In order to get honey to pass through these fine filters it must be heated slightly (to about 35°C). Once filtered the honey should be kept warm for a few hours in a settling tank to allow entrapped air bubbles to flow to the surface. After this time the honey will be ready for bottling.

Bottling

Honey can be bottled straight from the extractor or from storage. In both cases it can be coarse or fine filtered. Many people believe that coarse filtered honey (that containing the maximum amount of pollen but with bits of wax etc. removed) can help to reduce the symptoms of hay fever and prefer their honey to be in this state. Others like a bright clear golden liquid and yet others enjoy their honey creamy so that it can be spread thickly on toast. The state of honey in the jar is partly determined by the source of the nectar but also by the treatment of the honey by the beekeeper.

Honey can be bottled in any size of jar but looks best in glass jam or preserve jars with metal lids. Plastic containers can also be used and are available in many shapes and sizes, the 'bear'

Top: Passing recently extracted honey through a coarse filter.

Bottom: Essential jar filling kit.

jar being very common. When honey is bottled the jars must be completely clean. Even new jars can have specks of glass or other debris in them. Jars can be cleaned in hot, soapy water or by using a dishwasher on a hot cycle. Once cleaned the jars should be air-dried in an oven at a low setting before the honey is poured into them. Beekeeping suppliers sell buckets with taps that make filling jars a relatively simple procedure. The jars need to be filled to the rim with honey to reduce the amount of air trapped in the jar.

Finally a lid should be placed on the jar. If a plastic lid is used it must be cleaned and sterilised before use. Honey is corrosive and will rust a metal lid. If a metal lid is used it should be new rather than a recycled old one. The process of screwing a lid on a jar damages the coating on the metal and the lid will soon start to rust and potentially damage the honey.

It is important to remember that honey is a human foodstuff and at all stages the honey must be protected from contamination. All the utensils and containers should either be stainless steel or food grade plastic or glass. These should be cleaned and preferably washed in boiling water before coming into contact with the honey.

Honey types

It is quite possible to produce honey in three states depending on how you (or your friends and customers) like honey and also on the flowers bees have visited.

- **Clear**

  Clear honey is produced by passing the honey through a fine filter cloth. The honey is usually warmed to about 35°C and then passed through the cloth. The resulting liquid should be kept warm for about 24 hrs and then bottled. The process will remove most crystals of sugar and bubbles in the honey and leave a bright clear liquid honey. This honey may keep clear for many months and can usually be brought back to this state by gentle warming should it start to granulate in the bottle. Unfortunately warming can eventually degrade the honey (destroying the flavonoids

Reusing jars

In some counties the re-use of glass jars for bottling honey is considered inappropriate.

However it is accepted in the UK that small scale producers can re-use honey jars but that they are responsible for ensuring that the jars are clean and sterile before bottling.

Honey should always be bottled in a clean environment to reduce the risk of any contamination.

Using a refractometer to check the water content of a small sample of honey. A beekeeper will do this before starting to clear a super and taking the frames away for extracting.

and enzymes in the honey). Honey that is high in glucose will granulate faster than honey that is high in fructose. If the bees collect nectar from plants that produce high glucose honey it is probably better not to try to produce clear honey as it will rapidly revert to set honey.

■ Set

Set honey is honey that has been jarred as a liquid and then left to granulate naturally. The sugar crystal size and therefore the 'graininess' apparent when eating the honey depends on how fast the honey granulates. High glucose honeys will granulate and therefore appear 'smoother' to taste. The rate of granulation is affected by the temperature at which the honey is kept. If the temperature is about 14°C, maximum rate of setting is achieved; whereas below 10°C honey becomes very stable and does not granulate further.

■ Soft set

Many people like honey to be the consistency of butter so that it can be spread thickly on toast! Virtually any honey can be used to produce soft set honey. The honey is warmed until liquid and allowed to cool to room temperature. About

Above: Judge at a honey show checking a piece of cut comb for quality and taste.

Right: National Honey show competitors honey to be judged.

10per cent of soft set honey is then added to the liquid as a 'seed'. The seeding honey can either be previously produced soft set honey or a warmed jar of oil seed rape honey (which always granulates quickly to produce fine crystals). The liquid honey and the seed honey are mixed thoroughly without introducing any air bubbles (stirrers can be bought from bee keeping suppliers). Once thoroughly mixed the honey can be bottled and left at about 14°C until set. If the honey is kept in a cool place it will remain as soft set for a long time.

Two jars of honey ready for labelling.

Selling your honey

There are a number of laws regarding the sale of honey that ensure the consumer receives a safe and high quality product. Even if your honey is not sold it is worth preparing any honey that you might give away to friends or for your own use to the same standards.

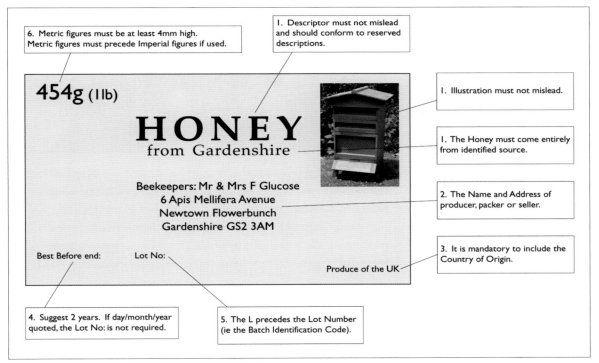

6. Metric figures must be at least 4mm high. Metric figures must precede Imperial figures if used.

I. Descriptor must not mislead and should conform to reserved descriptions.

454g (1 lb)

HONEY
from Gardenshire

Beekeepers: Mr & Mrs F Glucose
6 Apis Mellifera Avenue
Newtown Flowerbunch
Gardenshire GS2 3AM

Best Before end: Lot No:

Produce of the UK

I. Illustration must not mislead.

I. The Honey must come entirely from identified source.

2. The Name and Address of producer, packer or seller.

3. It is mandatory to include the Country of Origin.

4. Suggest 2 years. If day/month/year quoted, the Lot No: is not required.

5. The L precedes the Lot Number (ie the Batch Identification Code).

The first requirement is that the product is a true honey and not a substance that has been damaged by overheating. Honey should normally have no more than 21 per cent water in its

Example of a label that is accurate with correct size of weights and best before date.

content, the exceptions are specific honey floral sources mentioned below. The water content is measured using a honey refractometer. Refractometers for honey cost from £40 to £80 but can be borrowed from other beekeepers. The limit is set because honey with higher water content may begin to ferment and produce alcohol and carbon dioxide. The honey may have an unpleasant taste and the jar can potentially explode. Honey normally comprises between 17 per cent water and 20 per cent water. There are some exceptions such as Borage honey that can contain as much as 22 per cent water and Heather honey that can be up to 23 per cent.

Above: A sample of the authors 'light honey'.

Right: A selection of honey from light to dark colours together with a jar of soft-set honey and fully capped frame of honey ready for extracting.

From the time it leaves the hive to the time it is safely bottled, the honey must not be contaminated with any extraneous material. Contaminants can include bits of glass from the jar, human or animal hair and other debris introduced during processing such as earth or grass picked up while you are transporting the frames from the hive to the extracting room.

Honey should also be filtered to remove any bits of wax or other debris from the hive such as bees' legs, but should not be filtered with a mesh so fine that it removes all the pollen naturally introduced by the bees.

Having poured clean and wholesome honey into the jar it is also a requirement that it is labelled to conform with legislation.

The laws on labelling vary from time to time but in essence all honey must have a label that takes account of the following considerations.

- It is not misleading (honey collected from urban gardens should not have a picture of a heather moor on it).
- It is not incorrectly described (honey cannot be labelled as organic unless all the flora in the range of the bees is certified as organic and also the bees are kept to organic standards).
- The weight of honey in the jar must be displayed. (Current EU legislation allows honey to be sold in any quantity).
- The country of origin.
- The address of the producer or packer.
- A lot number unless all the honey produced is considered a single lot.
- A best before date.

It is advisable to consult the appropriate legislation before embarking on selling honey commercially.

Honey labels

With computers and home printers it is possible to produce your own labels that look attractive. Using self-adhesive labels makes the job of applying the label to the jars simple.

However it is possible to purchase ready-made labels from beekeeping suppliers that conform to all the labelling legislation. It can be sensible to start with ready-made labels and then move to your own design once your honey production increases!

Award winners at the Royal Show when held in Stoneleigh Park, Warwickshire.

Getting started

Don't go out and buy a beehive and bees without first learning how to look after them. This would be a sure route to wasting money, getting stung and possibly losing all of your bees within a year. It is far better to learn a bit about bees and beekeeping before making any investment.

When considering taking up beekeeping, you need to remember that you are looking after a colony of animals and, although their lifestyle is rather alien to us, they deserve to be respected and treated with care.

It is difficult for us to know if a colony of bees is sentient (in possession of 'feelings') but it can show a remarkable ability to protect itself and respond to external stimulae.

Some understanding of the life of the honey bee and the colony are essential in helping you to look after this fascinating animal. Remember that you are not on your own: there are probably more than 30,000 beekeepers in the UK and many of them will be willing to share their experience of beekeeping and help prospective beekeepers to learn the craft.

This, the last chapter of this introductory book, will help you to start the practice of beekeeping by using all the help and support available through many beekeeping clubs, books and on the internet.

Bee stings

It is inevitable that you will get stung when keeping bees – whether you are a beginner or an expert. Honey bees are not aggressive, but if you make mistakes when handling them, it may induce a defensive reaction. The bee's sting has evolved to be a defensive weapon largely because the contents of a bees' nest are very attractive to many other animals. Honey is a very valuable food source, as are the larvae in the brood nest. Honey bee colonies have always been subject to attack and the sting has developed to be an effective defence weapon.

The venom

When a bee stings you, the initial sensation will feel similar to being pricked by a needle. The bee injects a small quantity of venom that disrupts the local area by damaging cells. What happens then depends on the reaction of your immune system. People who have never been stung before have only a mild and very local reaction. Once the immune system has experienced a bee sting, antibodies are produced so that with subsequent stings the body is able to react. The levels of reaction after the initial stabbing pain are as follows:

Mild reaction

- Minor pain
- Localised swelling and itching lasting for no more than a day or two.

Inconvenient reaction

- Significant swelling near the site lasting a few days.

A problem reaction – consult a doctor soon

- Generalised body reaction (nettle rash in parts of the body remote from the sting).
- Temporary shortness of breath and sweating.

Bee stings

About 20 per cent of beekeepers will show some reaction to bee stings apart from the initial pain at the site of the sting. This can vary from a mild swelling to a severe anaphylactic shock.

Fortunately the number who suffer shock are very rare and most beekeepers who experience this level of reaction will give up keeping bees and keep well away from honey bees in future.

While deaths from bee stings do occur, they are very rare and are by no means confined to beekeepers.

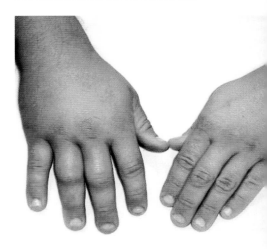

Typical localised swelling associated with a bee sting.

Be prepared

No matter how experienced you may be with honey bees and even if you have never had a problem with bee stings, it is never advisable to forego protection when near a colony of bees.

It is particularly important to protect your head, neck and body: being stung on the ankles is particularly painful.

Many beekeepers find that being stung on the hands is less of a problem than being stung on other parts of the body.

Always wear a full bee suit and wellington boots or similar when in the apiary.

If you find after a time that your normal reaction to bee stings is more than a localised swelling talk to your GP and they may advise the use of antihistamine tablets or that you always carry an 'Epipen' to be used if your reaction is dramatic and you suffer an anaphylactic shock.

Severe reaction, get help immediately

- Difficulty in breathing.
- Collapse.

If the person who has been stung suffers any reaction greater than swelling, a qualified first aider or doctor should be consulted. In the case of the last two, and most severe, reactions 'difficulty in breathing' and 'collapse' the emergency services should be called. The person will need immediate treatment from a paramedic and to be taken to hospital for monitoring and further treatment.

As soon as a person is stung they should be quickly removed from the apiary (to avoid further stings) and encouraged to relax. Observe carefully to discern the level of reaction and follow the advice above.

Learning beekeeping

This book can be used as a student manual to go along with an introductory course in beekeeping. This and similar courses are run every year by local beekeeping groups to help those new to beekeeping learn about the craft. These groups give their time and expertise to new beekeepers while the courses provide a basic grounding in the theory before beginners go on to handle bees. It may be that despite your interest and enthusiasm, beekeeping is not for you and attending one of these courses should help you to avoid finding out the hard way after investing in the equipment and losing your bees.

Reading

Over time, most beekeepers acquire their own personal 'library' of beekeeping books. There are very many fine books to choose from, either written by classic masters of beekeeping or by researchers on specific aspects of the craft.

It is easy to spend a fortune on books but if you join a beekeeping group you will usually find they have a library of the main titles and you will have an opportunity to browse some of the texts before buying them.

The internet is an excellent source of information on beekeeping and most search engines will find millions of references to bees or beekeeping. As with all Internet searches always consider the source of the information you find. In some cases it may just be the experience of a single beekeeper as opposed to the outcome of years of experimentation and analysis from a group. After a while you will be able to identify a number of trusted sites that are useful and practical in the information and advice they give. Some useful sites are listed at the end of this book.

You will probably find that once you start to enjoy your bee-keeping you will want to know ever more about these wonderful insects and, like many beekeepers, will start on a lifetime of learning more. It does seem to be the case that the more you learn about bees and beekeeping, the more the bees seem to behave themselves and work with you whenever you need to do a specific manipulation!

Meetings, talks and courses

Apart from local beekeeping groups who provide an annual programme of lectures and demonstrations, regional groups run training and demonstration events. Beekeeping associations also hold many short courses and conventions around the country where expert beekeepers and scientists will talk about various management techniques and the latest research on all aspects of beekeeping. These are low cost events and it is very worthwhile attending them.

Practical tuition

Beekeeping is essentially a practical activity and, as with any form of animal husbandry, it requires an understanding of how colonies of bees live and react to external stimulae. Books and courses are very important but until you have spent time working with a colony of bees under the watchful eye of a mentor, it is very difficult to know if beekeeping is for you.

When you start to keep bees, support from an experienced local person is vital. Bees do not always do as expected and knowing why they are doing something can sometimes be difficult. Having a good mentor makes all the difference. Try to go to as

Conferences

At the national level, beekeeping associations such as the British Beekeepers' Association (BBKA) hold courses and conventions on beekeeping. The main national events in the UK are the BBKA's Spring Convention, usually held in April which includes an international lecture programme, many workshops and seminars and a one-day trade show.

The other national event is the National Honey Show, held in late autumn. This also has a lecture programme, workshops and a trade show. However, this event has an international Honey Show at its core in which beekeepers compete in over 150 classes from honey and wax to photography and inventions and a whole host of other aspects of beekeeping.

Apimondia is a the biennial international conference, specifically dedicated to bees and beekeeping. At this conference, over 4000 beekeepers and bee scientists gather to listen to lectures and exchange views. It is associated with an international trade show of beekeeping equipment.

Unlike many other similar disciplines, beekeeping is an activity where, unusually, research scientists and amateur beekeepers can gather together to exchange views and to learn from each other.

Assessments

The BBKA and other National Beekeeping Associations run a national examination and assessment system so that having learned an aspect of beekeeping you can check to see if you have met a fixed standard.

These assessments are not compulsory but do show a commitment to beekeeping and successfully completing them will identify a beekeeper as being able and competent.

A list of examination grades and past papers are available from the BBKA web-site.

These examinations lead to becoming a qualified beekeeper. The highest level offered by National Associations is the Master Beekeeper or equivalent.

There is just one qualification above this level offered at International level – the National Diploma in Beekeeping (NDB) This examination is at the highest level and most beekeepers qualified at this level, become lecturers, tutors and demonstrators, helping others reach these heights.

many beekeeping meetings as possible when you start. It can be confusing because individual beekeepers may have very different ways of tackling a particular situation. They may all be right or all wrong but you can only assess this by watching how the bees respond. Even if you have had 25 years of beekeeping experience it is possible to learn from watching other beekeepers demonstrating. Many beekeepers spend their first year going to meetings and lectures and then decide that they need know no more. These people often find keeping bees very difficult and after a couple of years, when their bees die off in winter, they give up.

Different interpretations

Although many aspects of beekeeping can seem very confusing when you first start, you will find most experienced beekeepers will be delighted to explain the finer points of the craft whenever they are asked. Ask the same question of a number of different beekeepers until the answer makes sense to you. You'll find that some beekeepers understand fully the theories behind beekeeping, while others are simply good practical beekeepers.

Unfortunately there are also beekeepers who have learned a way to look after their bees, but have not fully understood the theory and have never tried different ways of keeping their bees. These beekeepers may be very helpful when you are in the first stages of your beekeeping experience but you will soon need to seek out others. These others will be people who do understand bees and have tried many ways to manage them and come to conclusions that recognise that the treatment of bees does not depend only on what you have been taught but also on the specific state of the bees. This can vary from day to day and it is crucial that you are able to read the state of the bees in order to know how to work with them. This is one of the great joys of beekeeping.

Many beekeeping groups make their demonstration and talks programme a social event, visiting their members' apiaries. This relaxed environment can help to develop new friendships and make it much easier for new beekeepers to ask those who have more experience for their help and advice.

Setting up an apiary

Choosing where to place your bees can sometimes take time. Most new beekeepers like to keep their bees in the garden, if it is big enough and there are no safety concerns. When selecting a site for a colony of bees, it is of primary importance that it will not be a nuisance or hazard to you and your neighbours.

Neighbours

Many people are frightened of bees and the thought of 60,000 bees just over the hedge is too much for some neighbours. It is better not to get into an argument that might sour relationships for many years. Talk to your neighbours and explain what you are proposing to do. If they have any concerns try to explain that bees are only defensive near their hive and then only when the beekeeper is working them. Offer the prospect of some jars of honey as these always go down well. If your neighbours remain a bit doubtful find another place to keep the bees and leave the idea of bringing them to the garden for another time.

Family

Bees in gardens can cause three problems for your family:

- Bees defecate on the wing. If they are flying past your washing line then there will be small, brown spots (bee poo) on the washing. Careful positioning of the hive and washing line can minimise this but it is difficult to eliminate

- When they are foraging the bees' flight from the hive tends to follow a low trajectory, as this is the most economical use of energy. Eventually the cruising height of about five metres is reached and the bees will be above any human physical space. If the hive is sited near an area that your family pass or spend time in the garden the bees may sometimes bang into people and cause some consternation.

To stop this the hive can be placed behind a hedge or fence that is two metres high. The bees then have to fly up quickly to get over the hedge and will not cause a problem in the rest of the garden. Interestingly when the

Keep out – honey extraction in progress! The bees are keen to have their honey back, so have collected on the outside of the beekeeper's shed door, whilst extraction is in progress.

195

bees descend to forage on flowers in the garden they present no dangers and meekly get out of the way of any human or animal presence

- Bees can be upset by animals coming too close to the hive entrance. Dogs and cats will soon learn to keep away but there have been occasions when they have been attacked by the bees. If you keep chickens and other similar animals it is wise to fence them off from the area surrounding the hive

None of these problems are insurmountable and one of the greatest delights of keeping bees in the garden is in watching them at the hive entrance, noting the pollen they bring in and assessing how busy the colony is from the activity at the entrance. In the garden it is much easier to visit the hive whenever there is a spare moment. Many beekeepers find it much more interesting than watching TV and even put a web-cam near the hive entrance!

Flat roofs

Of course, bees do not need to be kept in a garden. In town centres, hives are often kept on flat roofs above ground. This has the advantage that the bees are out of the way of other users of the garden. The main problem with keeping bees above ground is that without a safety fence it is possible for the hive (and potentially the beekeeper) to fall off. Hives get heavy when full of honey and if the roof is not strong enough, the bees and honey may end up in your house. Despite the dangers there are many examples of beekeepers having hives on roofs or similar 'out of the way' places.

Hives are kept on a number of high-rise buildings but it has been found that if bees are kept above about 15 floors they do not thrive. This is because they have to use too much energy flying up to the hive after foraging at street level and there is too little nectar left for the colony. If you have nowhere else to keep bees but are very keen to do so it is worth thinking about any strong flat roofs that might be available to you.

A well thought out 'out-apiary' with shade from the mid-day summer sun from the trees bordering the site.

'Out' apiaries

Keeping your bees away from your home is the only option for many beekeepers and is always worth considering. Many people with large gardens may be happy to offer a place for a beehive so that they have plenty of bees in their garden to pollinate their plants. Farmers are often happy to allow a beekeeper to use a corner of a field to hold a number of hives.

Allotments

Allotment societies are sometime interested in asking a beekeeper to keep a hive or two on the site to pollinate the fruit and vegetable crops. In these cases you need to ensure the bees are not a nuisance to the allotment holders; they usually are kept in a corner of the allotment site behind a 2 metre hedge.

If you are asked to place a colony of bees on an allotment site, it is wise to ensure that all the allotment holders are happy about having bees near their patch. Whilst all site owners want their crops pollinated some may be frightened by bees or even allergic to stings. Despite having permission from the land-owner or local council it will not be advisable to create a situation where the slightest problem with insects is blamed on your bees.

Most site owners will require the colony owner to have insurance cover in the case of any accident or harm caused to the allotment holders. Being a member of a local beekeeping group affiliated to the BBKA will ensure that you have sufficient cover and access to information regarding advice on siting apiaries.

Allotments are, by their very nature public places so your beehives may be potentially at greater risk of vandalism. It is not advisable to leave any spare equipment nearby unless there is a lockable and secure shed available. The hive components should be strapped together so that if the hive is tipped over it does not fall apart and expose the bees.

When considering an out apiary the needs of the bees should be considered alongside your needs as a beekeeper. The following are important to get right so that your beekeeping can be enjoyable:

A magnificent row of runner beans in an allotment. Whilst honey bees are not able to feed on these flowers due to their relatively short tongues, many other pollinators do.

The author's garden apiary – well away from the house and screened from neighbours and the road. The bees will provide excellent pollination of the nearby fruit trees.

A site **close to where you live**. If you need to drive many miles to the site it may inhibit you from visiting regularly and the costs of transport can escalate, particularly if you do not have a car! A vehicle is not essential but you need to carry some equipment with you, including your bee suit. A bicycle with a carry frame will suffice but when the time comes to carry supers back and forward this can become difficult and you may need to ask a friendly car driver to help.

Away from other beehives. In any area there is a limit to the number of colonies that can live and thrive and this is controlled by the quantity and quality of the local forage. There is also always a danger of disease spreading if apiaries are too close together. If you find a suitable site it is worth looking around and asking if there are any other local beekeepers. If there are and they are unhappy about you establishing an apiary it may be better to look for another site.

Hidden from the public. It is strange, but beehives can attract individuals who seem to think it is brave to upset the hive. Also, if a hive can be seen from a public thoroughfare, an unscrupulous beekeeper might try to steal it and its contents. There are people who steal hives and then sell them on to unsuspecting beekeepers or extract the honey for their own use. If the site can be seen from a road or footpath it helps if the hive parts are marked (branded with your post code) and camouflaged by painting them a neutral colour. It also helps to ensure that access gates can be locked if possible.

On level ground. Beehives are much easier to handle if they are placed on level ground and the surrounding ground is level. It makes inspecting the hive and placing components on the upturned roof much safer. Carrying heavy equipment to and from the apiary site is easier if the ground is flat.

Water nearby. Bees need water to dilute their honey and to air-condition the hive to control the temperature. They are not fussy about the dirtiness of the water and will happily drink from muddy pools or off marshy land. The location of the hive should be near a constant source of water where the bees can collect it without drowning. Cattle troughs are not appropriate and will result in many drowned bees. If there is no natural source of water nearby you can provide a receptacle and ensure that it always has water in it. An upturned dust-bin lid with stones, earth and moss in it, filled with water provides a suitable source.

- **Protected from cattle and sheep**. Beehives can be attractive to cattle and sheep who will delight in rubbing against them and possibly knock them over. If a farmer gives you a corner of a field it is always worthwhile putting stock fencing across the corner. Even though there may not be any stock in the field it also creates a barrier to prevent walkers, dogs or even farm workers and tractors getting too close to the hives

- **Early sun.** It has been shown that beehives that catch the early sun but perhaps are partially shaded during mid day do better than those in full sun or full shade all day. It helps the bees to warm the hive in the morning and commence foraging as soon as possible

- **Plenty of room around the hives**. When siting the hives in an apiary remember that you will need space between the hives and in the enclosure so that you can inspect the hives. There is always a need for space to put hive components on the upturned roofs and there are occasions when you will need to set up another hive or nucleus box to control the bees' swarming impulse. It is ideal if there is at least 1 metre around each hive

- **Vehicle access**. When the time comes to remove honey from the hive or to bring another hive to the site it will help considerably if you can drive a close as possible to the apiary site. Driving across a ploughed field is not an option but if the corner of the field is near the access track it makes life a lot easier

A sturdy fence protects an apiary from nosey cattle.

Wherever you site your apiary you should bear in mind that over time you may need to house more bees at the site but, at the same time, you need to enclose the area. Try to make as large an enclosure as possible and, if this site is remote from where you live, consider putting a small shed on the site to house spare

Using a swarm

Many beekeepers see a swarm as a cheap way to get a new colony. While this may be fine for an experienced beekeeper, there are a number of dangers for the new beekeeper.

The problem with a swarm is that often you have no idea where it came from or the state of the parent colony.

Swarms can be a source of disease, which is not immediately apparent. The general advice on collecting swarms is to keep them away from other colonies until the disease status has been checked.

Secondly, a swarm will invariably have an old queen with it and sometime later in the season this queen will be superseded, again this may be fine for the experienced beekeeper but as a new beekeeper it is better to start with a young queen that should see you through the first year of beekeeping.

Many beekeeping associations work as a group to collect swarms, re-queen them with a new queen and then allow the colony to produce brood on new frames before checking for disease. On the basis that all is well, this new colony can be offered to new members of the association as a starter nucleus colony.

equipment. This approach needs to be balanced against the possibility that vandals or bee rustlers who may wish to remove or destroy your equipment. Discretion is the key word and ensuring that prying eyes do not notice your apiary helps. Apiaries in your own or someone else's garden are safer but they are not an option for many beekeepers.

How to acquire bees

Having decided to learn about beekeeping and acquired your beekeeping equipment, the time will come to get your own colony of bees. You should never buy a colony in the autumn or winter. This is a stressful time for the bees and if they have not been looked after properly and prepared for winter they may die before the colony has a chance to get established. It is better to buy or acquire a colony in spring or early summer when the colony is quite small. It can then develop through your first summer as you become more confident in managing your bees.

Qualities of a colony

When you are starting beekeeping it is important to ensure that your first colony is small, docile and disease-free. In this way you will not be battling to keep the colony alive and nor will you become frightened of the bees. A large colony can seem initially daunting as many thousands of bees will be flying about while you are inspecting them. This is made worse if the colony is over-defensive and tries to defend itself.

In the spring, beekeepers and beekeeping suppliers prepare nucleus colonies for sale. Bees can be bought off the Internet or from beekeeping suppliers but if the price is very low you should be suspicious as there is probably something wrong with the colony. You need to find a reputable supplier and get advice from colleague beekeepers in your area. In general, colonies developed in your local area will be better suited to the particular conditions and it is worth talking to colleagues in your local association to see if there are any suppliers who can provide a nucleus colony for you.

Many beekeeping associations work to provide colonies for their new members. The advantage of acquiring bees this way is that if anything goes wrong you will have redress through the local beekeeping group and the bees should have been provided with a new young queen and the colony tested for disease.

The BBKA used to have a British Standard (BS1372) for a nucleus colony for supply; unfortunately this standard has lapsed but there is advice in the BBKA leaflet (L014 available on line) and from the National Bee Unit about the appropriate characteristics of a nucleus colony for sale.

It should have:

- A good-quality, young queen, probably marked with the appropriate colour for the year and possibly clipped
- Been checked and certified free of disease
- At least four frames with bees on them all
- Brood on at least three frames
- About 7-10,000 adult bees
- At least one frame of food
- Clean combs in good condition

Many of these criteria may be difficult for a new beekeeper to assess so it is always wise to take a beekeeping mentor with you when buying bees. If the bees are bought over the Internet or an-other way that means they will be delivered by post; you should ensure that the nucleus you buy will conform to the above speci-fications before parting with your money.

A colony with these characteristics should develop rapidly in its first season and by summer could be fully-sized and may even produce a little excess honey. The first year with a nucleus colony should always be considered as the time when the colony is given every opportunity to enlarge and become robust enough to survive the forthcoming winter.

Making a good start with two polystyrene nuc boxes and a brand new brood box ready to take the contents of one of them.

How much to pay

The price of a nucleus colony can vary enormously from year to year and location to location and it is very difficult to give guide figures. In 2014 the price of colonies varied from about £80 to £300. Most come with a young queen but this cannot be guaran-teed. Price is no guide to the quality of the nucleus but a colony costing less than £130 does not represent true value for the work

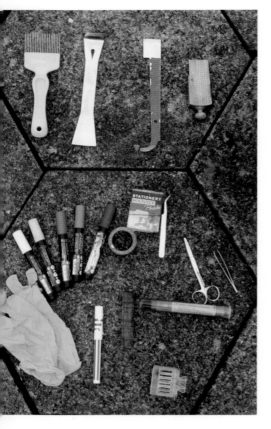

A selection of essential tools.
Top row left to right: uncapping fork, spade type and J type hive tools and wedge.

Middle row: queen marker pens, queen marking cage (crown of thorns), drawing pins, curved tweezers, scissors, small fine straight tweezers.

Bottom row: nitrile gloves, pen, wire queen cage, Epipen and bottom, another style of queen cage.

involved in creating a nucleus and is likely to be lacking at least one of the criteria for a good nucleus.

The exception to this is to purchase a nucleus colony from your local beekeeping association where the group may subsidise the cost by providing labour freely or as part of the package given to their new members. Honey bee colonies are not cheap but if they are cared for they will last many years, with the occasional natural re-queening and replacement of the brood frames when they become old and badly stained.

Basic equipment requirements

When you start beekeeping there is a wide array of equipment that you could buy and it is very tempting to get 'fully equipped' just in case you might need this or that tool in the near future. The list of essential equipment is short, along with a longer list of useful items. A great deal of beekeeping equipment can be bought second-hand and, as long as it is in good condition, it will give you many years of service and reduce the cost of keeping bees. If you go to a beekeeping auction or want to buy second-hand, try to have assistance from an experienced beekeeper from your local association.

Although there are bargains to be found, sometimes you may be asked to pay a high price for something that has reached the end of its useful life. If you are buying hive components there can be a big difference in price between those made of western red cedar and those made of other woods. This wood is ideal for beehives in that it does not warp and does not rot.

Hive tools can vary greatly in price and if you can afford it, stainless steel has advantages over mild steel. It can be left in a bucket of washing soda solution to remove debris from a hive without rusting. Painted tools are fine but may not last as long.

Essential equipment

- **A bee suit with veil.** This can either be a jacket or full-length suit. If you choose a jacket make sure that the waistband fits snugly otherwise you might find a few bees walking up your back inside the jacket.

- **Wellington boots.** Most of us will have these already. They help in that they will stop bees walking up inside your trousers or, in the worst case stinging your ankles – a particularly painful place to be stung.

- **Gloves.** Some beekeepers prefer not to wear gloves but if you don't you may pass your pheromones onto the bees and cause some upset. Gloves are not really designed to prevent bees from stinging you but if you are stung, they are a simple way of removing the sting from your hand. Leather gloves harbour diseases and are extremely difficult to clean. Rubber washing-up gloves are relatively cheap, easy-to-clean and provide a thin barrier between your hands and the bees. Disposable non-sterile latex gloves or nitrile gloves are equally fine and have the added advantage that there is no loss of feeling when wearing them. Many new beekeepers feel that there is a bit of protection from being stung by using washing-up gloves but when you feel ready you should try to use the thinner disposable gloves. However, you should remember that bees will sting you if you crush them or upset the colony when inspecting it. If you treat your bees with care and consideration they are far less likely to try to sting you.

- **A hive tool.** These come in many shapes and sizes. There are two primary types, the 'J' type and the spade type. The first is particularly useful if you have a beehive that has frames with short lugs. The spade type is probably more useful if your frames have long lugs. Some bee keepers have both types but you should probably buy your own hive tool once you have decided on the type of hive that you will use. It is a good idea to borrow hive tools from your colleague beekeepers and see which type is most comfortable for you. Bear in mind that you should be able to do virtually all your beekeeping manipulations without the need to put your hive tool down.

A good quality stainless steel hive tool with a thin spade end will easily prize open hive parts without causing damage to the brood and super boxes.

- **Wedge.** A wedge is used to hold one of the boxes away from the lower one when aligning to hive. It helps to

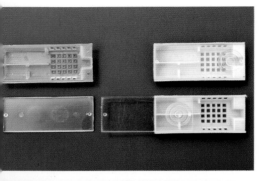

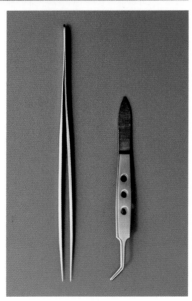

Top: basic queen cages
Middle: record keeping as hard copy, on a laptop computer and on a mobile phone.
Bottom: two alternative styles of fine tweezers.

ensure that bees are not crushed whilst assembling the hive. Not all beekeepers use wedges but they are quite helpful when starting beekeeping.

- **Uncapping fork.** An uncapping fork is used when inspecting drone brood for the presence of Varroa mites. It can also be used for removing cappings from sealed honey so that the honey can be extracted.

- **Smoker.** A smoker (see page 66) is a very useful piece of equipment and even those who like to inspect their bees without the use of smoke are advised to have one nearby in case the bees suddenly react badly to the inspection. When you buy a smoker, try to obtain one that is made of stainless steel and quite large. A new one like this will cost over £30 but will give good service for many years.

- **Queen cage**. Even as a new beekeeper it will help some manipulations if, having found the queen you can put her in a cage to protect her while you are working. These come in many shapes and sizes but only cost a pound or two. Watch your fellow beekeepers and try the cages they have to find which you find easiest to use.

- **Record book.** Any book or set of sheets will do. When you begin it is a good idea to follow the suggestions of others on what you need to record. Always look at your record book before inspecting the bees; it will remind you of what happened last time and what you should find and do this time. Fill out the book as soon as you finish inspecting the hive – it is too easy to forget what you did or think should be done next time. At some point in the future you will have more than one hive and at that time immediate recording will become essential; getting into good habits now will help later on. It is now possible to use a recorder or phone app. and then download records onto a computer spreadsheet at a later date.

- **Fine tweezers.** There will be occasions when you want to look at a particular larva or remove a piece of wax that is obstructing your view. A pair of straight tweezers with small flat ends will be very helpful.

- **A hive.** You will need to decide the type of hive you want. From the bees' point of view it needs to be sufficiently large, dry and have just one small entrance. When you buy a hive, ensure that you have all the components (stand, floor, brood chamber, queen excluder, two supers, crown board, and roof). All the components should all be clean and free of any wax debris. If it is bought second-hand you will need to clean every component and scorch the insides of the brood box and supers.

- **Frames and foundation.** These will be bought from a beekeeping supplier. Do not be tempted to buy second hand frames or foundation. You never know where they have been and they may contain disease pathogens. The frames are easy to make up and to fit with wax foundation. All you need are some frame nails, a sharp knife and a hammer.

How to build a colony

Colonies are often bought in 'travelling' nucleus boxes that are not designed to house them for any length of time. As soon as your colony arrives it should be given a couple of days in the position where the hive will stand so that the bees can acclimatise to the new surroundings. The colony should then be moved into your hive by carefully moving the nucleus to one side. Then take each frame out and place it into your hive. These should be kept in their same relative position and a frame of foundation placed each side of the colony. A dummy board is placed to one side of the frames to reduce the overall capacity of the hive.

The colony is then closed up and fed with a sugar solution to encourage the bees to draw out the frames of foundation.

After a week, inspect the colony and be prepared to add another two frames of foundation and feed again. Continue this process until all the brood frames have been drawn out and can be used by the bees to store honey or raise brood. Now is the time to

Spare equipment

If you start with a small nucleus colony in the spring or early summer it is unlikely that you will need any spare equipment but in your second year the following will be a must.

Brood box *A second brood box is helpful for controlling swarms. You will need a set of frames and foundation to fill the box.*

Another super *With a full sized colony in the spring you will hopefully have a good crop of honey (weather permitting).*

Nucleus box. *This is useful and can help with a number of manipulations. In particular it can be used to hold the first frame you take out of the brood box when inspecting the bees or the frame with the queen on it.*

Queen marking pen and crown of thorns *You will probably produce a new queen in your second year and it will help if you have a marking pen to mark this queen once she has settled down to lay eggs and has her own brood in the colony. The crown of thorns is a device used to trap the queen on the comb without damaging her so that she can be marked.*

A second floor, queen excluder, crown board and roof *If you intend to have a second colony then you will need equipment to house the colony. It is also helpful to have these items for swarm control.*

Set of queen marking pens arranged in the correct 'date' order left to right. 'Will You Rear Good Bees' is a useful way of remembering the sequence of colours.

Queen wing clipping scissors – a useful addition to the tool box.

consider placing a queen excluder over the brood box and adding a super. This time do not feed the colony, as there is a danger that your sugar solution will be stored by the bees and end up in your honey.

If you start with a nucleus colony in May it is quite possible that the colony will be almost at its full size by July and will provide a surplus of honey for you. If the nucleus colony arrives later than May it will build up but is unlikely to provide honey. It will need all it gathers for the forthcoming winter.

The most important action is to ensure that the nucleus colony you acquired will survive through the winter months and be strong enough to take advantage of the following summer. No matter when you acquire the swarm you must care for it in those first few months and give the colony every opportunity to build up to full-size. Your efforts will be rewarded in the following years.

Insurance and liabilities

Honey bees are legally classed as wild animals but if you choose to manage a colony then any resulting damage to people and property is your responsibility. Fortunately, should you join a local beekeeping group that is affiliated to the BBKA then you will be covered for public liability insurance to a sum of £10 million. Each year there is on average one claim against this insurance policy. Claims range from accidents following someone being stung to damage to a neighbour's pet. These incidents are very rare but it is reassuring to know that if something untoward does happen there is cover, should anyone sue you.

Ancient beekeeping laws decree that once you have decided to collect a swarm it becomes your property and that during the collection you are liable for any damage. In the case of swarm collection from a garden, always ensure that you have the owner's permission before starting and, if the collection requires you to clip a shrub or tree make sure that the owner understands and agrees.

A word of warning; be careful not to put yourself in any danger when managing your bees or collecting a swarm; the insurance described above covers only third parties and not the beekeeper.

This may seem a strange aspect of beekeeping to end this book but it should be seen as positive. The BBKA works hard to promote responsible beekeeping in care of the bees. It is also in recognition that we live in a litigious society and there are occasions when the knowledge of appropriate insurance is gratifying. Accidents rarely happen and sometimes the cause is not down to the beekeeper or even the bees. Having a legal framework governing the way we carry out our beekeeping means that there is an organisation ready to represent us if a claim is made and that we can rest easily at night.

Happy beekeeping

Taking the responsibility for keeping a colony of bees can seem quite daunting when you start. These insects are not domesticated and will respond naturally to the condition that face them. For many reasons honey bees are both fascinating and difficult to understand. After a number of years we can learn the basics and become adequate beekeepers but, perhaps, not fully understand the 'whys' behind the manipulations we do to help our bees prosper.

Our advice to you is to take every opportunity to talk to other, more experienced, beekeepers. Join the local beekeeping association and go to meetings and try to get to the national beekeeping events. There is much to learn and each time you learn something about bees and beekeeping you will find you become a truly life-enhancing insects.better beekeeper! We wish you well with this important and interesting hobby and hope you enjoy becoming one of the many people who maintain the craft of beekeeping across the world. Honey bees are probably the most researched insect in the world and, as a single species, have the greatest effect on humanity's wellbeing. New information coming from the research will help us all to become better beekeepers. We are all learning and can never consider that any of us knows it all. We hope you enjoy your beekeeping and learning more about these truly life-enhancing insects.

The author carrying out a quick autumn inspection.

■ Useful contacts

Government Agency

The National Bee Unit
www.nationalbeeunit.com

Beekeeping Associations

The British Beekeepers'
Association
T: 02476 696679
www.bbka.org.uk

The Scottish Beekeepers'
Association
www.scottishbeekeepers.
org.uk

The Welsh Beekeepers'
Association
www.wbka.com

The Ulster Beekeepers'
Association
www.ubka.org

The Institute of Northern
Ireland Beekeepers
www.inibeekeepers.com

The Federation of Irish
Beekeepers' Associations
www.irishbeekeeping.ie

Bee Improvement and Bee
Breeders' Association (BIBBA)
www.bibba.org

National Diploma in
Beekeeping (NDB) www.
national-diploma-bees.org.uk

International Bee Research
Association (IBRA)
www.ibra.org.uk

Central Association of Beekeepers
www.cabk.org.uk

Equipment Suppliers

Agri-Nova Technology Ltd
www.agri-nova.biz

Apimaye UK
www.apimaye.co.uk

Arnia
www.arnia.co.uk

BeeBay
www.bee-bay.net

BB Wear
www.bbwear.co.uk

Bee Basic
www.beebasic.co.uk

B J Sherriff International
www.beesuits.com

Brunel Microscopes Ltd
www.brunelmicroscopes.co.uk

Chemicals Laif
www.chemicalslaif.it

C Wynne Jones
www.beesupplies.co.uk

Compak (South) Ltd
www.compaksouth.com

E H Thorne (Beehives) Ltd
www.thorne.co.uk

Fragile Planet
www.fragile-planet.co.uk

Freeman & Harding Ltd
www.freemanharding.co.uk

Maisemore Apiaries
www.beesonline.co.uk

Modern Beekeeping
www.modernbeekeeping.co.uk

National Bee Supplies
www.beekeeping.co.uk

Paynes Southdown Bee Farms
www.paynes-beefarm.com

Park Beekeeping Supplies
www.parkbeekeeping.com

Stamfordham Ltd
www.stamfordham.biz

Vita (Europe) Ltd
www.vita-europe.com

Bee Book Suppliers

C. Arden Bookseller
www.ardenbooks.co.uk

Bee Books Old & New
www.honeyshop.co.uk

International Bee Research
Association (IBRA)
www.ibra.org.uk

Northern Bee Books
www.beedata.com

Beekeeping Publications

BBKA News
www.bbka.org.uk

Bee Craft
www.bee-craft.com

Beekeepers Quarterly
www.beedata.com

Gwenynwyr Cymru
www.wbka.com

The Scottish Beekeeper
www.scottishbeekeepers.org.uk

Bee World
www.ibra.org.uk

Bee Improvement and Conservation
www.bibba.com

Glossary

Abdomen Rear segment of an insect. It usually contains the digestive, excretory and circulatory system as well as many other vital organs for survival.

Acarine Mite that lives in the large breathing tube of an adult honey bee. It lives on the haemolymph of the bee and reproduces in the tube.

AFB American Foulbrood is a spore forming bacterial infection of honey bee brood. It can be fatal to the colony and, once detected in the UK will mean the colony must be destroyed.

Anaphylaxis Medical condition of animals that results in collapse and a sudden drop in blood pressure. It can be fatal if not treated quickly.

Antenna Projections on the front of the head of a bee that are packed with sensory organs (primarily smell, taste and touch) and used by the bees to assess its surroundings.

Anthers Male part of a flower that produces the pollen.

Appointed Bee Inspector Government appointed person with the authority to inspect colonies of bees to identify any potential diseases. They have the authority to destroy colonies or prevent movement of bees suspected of having certain diseases or harbouring certain notifiable pests of bee colonies.

Bee bread Mixture of nectar and pollen stored in the hive and used to feed young larvae.

Beespace Gap of 6–8 mm through which a bee can crawl. This is an important dimension because in a hive bees will fill a gap smaller than this with propolis or wax while larger gaps are filled with wax comb.

Beetight Hive with no external gaps through which a bee or other similar insect can gain access to the interior except for the entrance to the colony.

Brood chamber Part of a beehive where the queen resides and brood is developed.

Brood food Collective name for the food that is provided to larvae by worker bees in the hive. Constituents vary according to the age, gender and cast of the larva and comprises pollen, nectar/honey and glandular secretions from the workers.

Bugonia Ancient belief that colonies of bees were spontaneously evolved from the carcasses of dead oxen.

Caste One of the distinct forms of animal found in social colonies. There are considered to be two castes in a colony of bees namely the queen and the worker. Both are female but have very different physical forms.

Chalk brood Fungal disease of bee larvae. The fungal spores infect the larvae and rapidly turn the larva into a white hard chalk like lump. The remains can turn black or grey when the fungus produces spores.

Chitin Material comprising the outer casing on a bee (and other insects). It is hard and waterproof and an organic polymer.

Clearer board Board the size of the external dimensions of the hive that allows the bees to move one way, usually downward. It can be used to remove bees from supers that are to be taken from the hive to extract honey. After a time it is possible to remove the supers with virtually no bees left in them. Often clearer boards are let on the top of the hive, just under the roof and act as a crown board.

Clipping a queen Process performed to prevent the queen from flying. It involves removal of about a third of one of her fore-wings with scissors. It does not seem to cause harm to the queen if done carefully. However, if she tries to fly she will spiral to the ground as the operation unbalances her. Some beekeepers believe this 'mutilation' is unnecessary. It should be done after the queen has completed her mating flights and is laying well. It should be done before the colony could consider swarming as one of the advantages of clipping a queen is that should the colony attempt to swarm it will return as it will not leave without the queen.

Cluster Term for a group of bees that gather together holding each others' bodies or wings with their legs. Bees do this for several reasons: to raise their body temperatures to produce wax; to maintain a warm temperature in the winter so that they do not chill and become torpid; and when swarming to provide a temporary place where they rest before departing to their new nest.

Copulatrix Properly known as the *bursa copulatrix*, this is the entrance to the vagina on the queen and is a cavity that is able to receive the endophallus of the drone and the ejaculate. It leads to the vagina and thus to the ovaries.

Crown board Board placed at the top of the hive to mark the limit of where the bees will travel. It often has a hole or number of holes in it to aid ventilation. The crown board is then covered by the roof that keeps the top of the hive waterproof and prevents top access by other bees and insects

Dances Bees communicate in many ways but one of the most spectacular are a series of stylised dances. These are usually performed on the wax comb inside the hive and the most well known are those that indicate the presence of food. The round dance just tells other bees that there is food near the hive and those observing the dance are invited to leave the hive in search of the food. The waggle dance is more complex and tells the observers that if they leave the hive in a certain direction and travel for a certain distance they will again find some food.

Dance floor Part of the wax comb in the brood chamber where bees gather either to dance or to watch and compute the meaning of the dance.

Driving the bees An old term that relates to a method for remove bees from one skep into another empty one. This allowed the beekeeper to remove honey from a skep without the prior need to kill all the bees.

Drone A male honey bee.

Drone collection area Geographical area where drones gather to await the arrival of a virgin queen requiring mating.

Dummy board This is normally a piece of wood shaped like a frame that is placed to the edge of the brood chamber. Its objective is to keep all the frames of beeswax with bees on held tightly together. It can easily be removed without damaging the bees to provide a gap in the hive to aid the removal of frames with bees on.

Eclose Hatch from an egg. The term is normally used when considering a larva that does not leave any shell as the egg case is consumed by the larva.

EFB European Foulbrood (EFB) is a bacterial infection of honey bee larvae. The bacterial lives in the larval gut and competes with the larva for food. The larva is very weakened or dies through starvation.

Extractor A piece of equipment used to remove liquid honey from honeycomb. It operates on the spin dryer principle whereby the frames are spun round and honey is pulled out of the frames by the centrifugal force.

Feeder A device that is put into a hive to allow the bees inside to gain access to a sugar solution to provide an alternative source of carbohydrate when there is not enough nectar available for the colony.

Feral Wild, or not domesticated.

Fertilisation Process whereby a female gamete and a male gamete join together to start a new life form. In the case of flowers this follows pollination and involves the pollen grain

growing a tube inside the stigma to the ovary of the plant.

Fondant Normally 'Baker's' fondant the white soft sugary paste placed on top of buns, and comprising sugar, water and glycerine. It is used to feed bees in cold weather.

Forager 'Older' bees in a colony that fly out from the hive to collect nectar, pollen, water and propolis.

Gamete Cell containing only half the DNA found in a normal cell. These are cells produced in the testes or ovaries that become sperm or eggs.

Ganglia Cluster of nerve cells along the length of the honey bee. These clusters provide local control to a part of the body of the bee.

General inspection Process of inspecting a colony of bees to assess the health and viability of the colony. General inspections are conducted from mid spring to early autumn and can be between one and two weeks apart.

Guard bee Honey bee whose duties are to remain at the entrance of the hive to ensure that only colony bees are allowed into the colony. Should a predator or a bee from another colony try to enter the guard bee will threaten the intruder and call up support from other bees in the colony.

Hackle Usually made of straw, this is a cover placed over a wicker or straw skep to prevent rain entering the skep.

Haemolymph Liquid in the body of the honey bee. It has a similar function to blood except that it does not deliver oxygen to the cells in the body. It does deliver nutrients and also keeps the body of the bee under pressure to maintain its shape.

Hefting Process to estimate the weight of a hive. One side is lifted just off the ground and the weight assessed. The weight of the hive and its components (including bees and stores) is twice the weight lifted. This gives a very approximate way of

assessing the stores in the hive without opening the hive.

Hive tool Instrument used to prise open components of the hive and lift frames from the boxes. Usually made of either stainless or painted steel.

HMF Abbreviation for Hydroxymethylfurfuraldehyde. This is a breakdown product of fructose in the presence of an acid and its rate of production is increased with temperature. It is used as an indicator of honey quality. Honey must have a level of 40mg/kg to be legal in Europe.

Honeydew Exudation from sap-sucking insects such as aphids. Bees collect this and the honey produced is called honeydew honey.

Honey stomach Also known as the honey sac. This is an expandable sack between the oesophagus and the stomach (ventriculus) of the honey bee. It holds nectar while the bee is foraging which can then be regurgitated and processed to form honey.

Hypopharyngeal gland Gland at the front of the head of a worker bee that can produce a component of brood food (known as clear substance) or, in foraging bees produces enzymes, to assist the processing of nectar into honey.

Instrumental insemination Mechanical means of inseminating a queen with semen harvested from selected drones. It is a specialist task that is normally only undertaken by queen breeders wishing to develop certain characteristics from their colonies or for research purposes.

Juvenile hormone Hormone produced in queen larvae and old adult workers. In the queen it seems to be the trigger for the production of a queen from a larva. In workers the levels of juvenile hormone increase as the workers age and appear to have an effect on the regulation of the relative numbers of house bees to foragers in any colony.

Larva Pre-pupal stage of an honey bee. Larvae grow rapidly from eggs and are fed by 'house bees'.

Lugs Appendages on the top bar of a wooden frame used to hang the frame in a component of a hive. Typically there are long and short lugs depending on the design of the hive.

Mandible All bees have two mandibles as part of their mouth. They are used to chew. Workers can use their mandibles to chew wax and make it pliable before using it to construct honeycomb. They also use their mandibles when defending the hive and attacking an intruder.

Mark a queen Process whereby the queen in a colony is made easier to find by painting a small dot on the top of her thorax. Coloured pens are used and there is an agreed colour for each of five years that is used in cycles.

Mating flight Queens mate on the wing and during their first three weeks of life must fly from the nest to mate with up to 20 drones. Her trips are known as mating flights.

Mono floral crop Description that notifies that the honey crop in a hive has come predominately from a single flower source.

Nasonov Gland at the rear of the abdomen of worker bees used to provide a pheromone scent that attracts other workers.

Natural beekeeping Recent practice that involves keeping bees with minimal disturbance by the beekeeper. Usually natural beekeeping would not involve the use of wax foundation and would not try to stop the colony from swarming. In many ways natural beekeeping is more complex than conventional beekeeping and if not done with care will result in either the loss of the colony or death from disease or pests. Conventional and natural beekeeping can exist side by side as long as there is no risk from transferrable diseases.

In urban areas natural beekeepers need to take special care that swarms of bees do not cause a nuisance.

Nectar Sweet substance produced by plants (normally from flowers) that is collected by bees and is the prime constituent of honey.

Nectaries Small protuberances on a plant, usually at the base of a flower that exude nectar. The nectaries produce nectar from the phloem within the plant.

Nest Part of a hive where the queen lays eggs and bees are bred. It includes parts of the hive where pollen and nectar is temporarily stored.

Newspaper method Process used to combine two colonies of bees to result in only one. Newspaper with slits is placed between the colonies on a single site. Over time the bees from each colony bite through the paper and intermingle. Without the delay caused by the paper the colonies could fight.

Nosema Fungal (micro-sporidial) disease of honey bees that attacks the lining of the ventriculus and reduces the bees ability to digest protein.

Nucleus box Half-size hive that is used to house a small colony while it expands to occupy a full sized hive.

Organic beekeeping This is a misnomer in the UK as there are very few areas where one can be sure that the bees will only forage on organic flora. There are other requirements to qualify for the term organic; namely the colony must not have been treated with medication for the last 12 months and the queen must not be marked or clipped.

Ovipositor Long tube at the rear of some insects that is used to deposit eggs either in rotten wood or another insect or larva. These are commonly seen on sawflies. In the honey bee the ovipositor has been modified into the sting mechanism.

Oxalic acid Organic acid used as a control for *Varroa*.

Pheromone Chemical produced by one animal to cause an effect on another animal of the same species.

Piping High frequency buzzing noise produced by a queen. It appears to announce to another queen that the queen is here and it is sometimes used by a queen to stop or slow an attack on the queen by workers.

Pollen Male gamete of a flower encased in a durable casing.

Pollination Process of moving a pollen grain to the stigma (female receptive organ) of a flower of the same species.

Polarised light Light waves where the orientation of the waves are in one plane only.

Polished queen cup A queen cup is the receptacle prepared by workers to receive an egg that will eventually become a queen. When polished by workers using propolis the cup is ready for an egg. Queens will not lay an egg in an unpolished queen cup.

Proboscis The drinking tube of a bee. Normally this is folded under the head but brought forward when the bees is sucking up a liquid.

Propolis The sticky coating on buds, particularly trees and shrubs, that is collected by bees to polish cells and seal up small gaps in the hive. It has antiseptic properties and can protect larvae from disease.

Proventriculus Valve between the honey stomach and the ventriculus (stomach) of the bee. It prevents nectar mixing with the contents of the ventriculus and has small hairs that filter pollen from the nectar to pass it into the ventriculus for digestion.

Pupa Stage in a bee's life cycle between being a larva and an adult. During this stage the cell is capped with a porous substance made from wax and pollen.

Pyloric valve Valve at the end of the ventriculus controlling the flow of food into the intestines.

■ **QMP** Queen Mandibular Pheromone. One of the main pheromones produced by the queen that is attractive to workers.

■ **Queen** Female bee in the colony that is able to lay large numbers of both fertilised and unfertilised eggs. The queen is differentiated from other workers by the way she is fed during the first three days as a larva.

■ **Queen cell** Wax cell specifically constructed to house a queen larva. This is usually large and hangs downwards (as opposed to horizontal) in the colony

■ **Queen cup** Start of a queen cell. They look a bit like acorn cups and are distributed through the brood nest ready for when the colony decides to make another queen.

■ **Queen excluder** Mesh that is placed between the brood chamber and the supers and has gaps big enough for the workers to pass through but not so large so the queen is confined to the brood chamber.

■ **Queen substance** Pheromone secreted by queens that prevents workers from being able to lay eggs.

■ **Read the bees** A process of observing the state of the colony when inspecting it that results in an ability to identify how the colony arrived at its current state and how it might progress in the future. It is a process that takes a lot of practical experience and tuition from good beekeepers.

■ **Record card** A place to note the current condition of the colony and hive so that on a subsequent visit the beekeeper is prepared for the next inspection. There are many designs of record card and some can be found on the Government's Bee Base web site or from the BBKA website.

■ **Refractometer** Device that determines the water content in honey by measuring the refractive index of the honey.

■ **Royal jelly** Collective name for the substance fed to queen larvae. It is almost entirely produced from the mandibular and hypopharyngeal glands of workers.

■ **Scout bees** Subset of bees in a hive that will seek new forage or, when required, a new nest site. It is not known if scout bees remain scouts all the time or if any bee can become a scout bee.

■ **Shaking bees** Method of removing virtually all the bees from a frame by shaking them into the hive. When done properly the bees are not harmed and does not result in thousands of bees flying about. Once the frame is clear of bees it is much easier to check the state of the brood.

■ **Skep** Hemispherical hive made of straw or wicker, used to house a colony of bees prior to the development of removable frame hives. These are often used as a convenient container in which to collect swarms or as an item of historical interest.

.■ **Social insects** Group of insect species that have developed a highly organised and cooperative lifestyle where a single (or very few) female will lay eggs, while the vast majority of other female are sterile and take on the duties of caring for the colony. Typical social insects are honey bees, wasps, ants and termites.

■ **Sororicide** Killing of a sister. In beekeeping this relates to a virgin queen killing her sister virgin queens in a colony to ensure only one survives to take on the responsibility of mating and laying eggs for the colony.

■ **Spermatheca** Gland within the abdomen of a queen that houses and nourishes all the sperm that she has collected on mating flights. The sperm are released a few at a time whenever the queen fertilises an egg to produce a worker or queen.

■ **Spiracle** Breathing tubes along the sides of a bee. Bees do not have lungs as such but breathing holes

leading to a mass of tracheoles that supply oxygen directly to the organs in the body.

■ **Sterile** State reached by a queen after about three to four weeks from hatching if she has been unable to mate with drones. After this period she is unable to be mated and can only lay the unfertilised eggs that result in drones.

■ **Stigma** Female part of a flower that is receptive to a pollen grain. Once a suitable grain attaches to the stigma the plant is considered to be pollinated and the grain will then grow a tube through the stigma to deliver the male gametes to the ovary of the flower to fertilise it.

■ **Strop** Process used by worker bees to regurgitate nectar onto their mouthparts and then to suck it back into the honey stomach. Stropping will reduce the water content of nectar to about 50 per cent at which point the nectar will be suspended in a cell to dry further.

■ **Sugar syrup** Liquid made up by the beekeeper or bought from suppliers that can be fed to bees when there is a danger that they could run out of food in the future. It is a substitute for nectar in the bees diet.

■ **Super** Hive component that houses frames that can be placed over the brood nest to provide additional space for the bees and a place to process and store honey.

■ **Superorganism** Term referring to a group of social insects that live and survive as a single colony. The management and organisation of the insects in the superorganism can often be seen as if the group is acting as a single animal.

■ **Supersedure** Process whereby a colony of bees are able to raise a new queen without swarming. The result can be either that the workers removed the old queen immediately or, in some cases, the old and young queen will remain side by side both laying eggs.

Swarm Group of bees that have left the original hive to set up a new colony.

Swarm control Method of managing the process of swarming so that the colony acts as if it has swarmed but in reality has not but has been divided by the beekeeper and housed in two hives.

Swarm prevention Mechanism for managing the colony so that the need to swarm is either minimised or delayed.

Thorax The middle part of a bee between the head and abdomen. This is where the legs and wings are attached and houses strong muscles used for flight and also generating heat.

Thymol Oil of thyme, usually artificially produced. It is the active ingredient in many medicines and treatments for honey bee colonies. It is particularly effective against *Varroa*.

Travelling nucleus box Box that will hold frames and used to transport small colonies. They are often used to collect swarms.

Trophallaxis Food sharing. A process much practised by bees in a colony for either transferring food from a foraging bee to a house bee or a method of building the cohesion between members of the same colony.

Uncapping Process of removing the wax cappings on supers of honey so that the honey may be extracted. There are many ways of doing this depending on the number of frames that need uncapping and the capital cost of the instrument.

Uncapping fork Devise specifically made for removing cappings from cells. It is fine if there are not too many frames that require uncapping but can be tedious otherwise an uncapping knife can be used. Uncapping forks are relatively cheap and most beekeepers will have one.

Varroa Parasitic mite of honey bees that originated on the Asian honey bee (*Apis cerana*) and jumped species in the 20th century. It is now the most significant pest to honey bee colonies and often results in the death of an untreated colony.

Ventriculus Stomach of a honey bee. It lies between the honey stomach and the intestine and is where pollen is processed to extract the proteins and other nutrients.

Virgin queen Adult queen that has yet to complete her mating flights. Virgin queens have not yet started to laying eggs.

Waggle dance Movement by one bee that passes information onto other bees to indicate the location of food for the colony. The waggle dance is done inside a hive and generally on an area specifically marked by the bees as a dance floor.

Winter bees Bees that, through adaptation and minimal work, are able to live for many months instead of weeks. Winter bees keep the colony alive when there are few or no sources of nectar and pollen. When spring emerges winter bees are able to raise the next generation of bees that will take the colony forward into summer.

Worker Female bee that does not lay eggs but will supports and provisions the nest where a queen lays all the eggs. Whilst a worker is not totally unable to lay eggs she is prevented from doing so by the presence of a queen and brood pheromones.

Young Bee Generic term, interchangeable with house bee, that indicated the bee will spend the majority of its time in the nest with duties relating to the wellbeing and management of the nest. Once a bee is about three weeks old it will become a forager (or old bee) where the duties are predominately that of providing stores for the colony.

Index